CAMBRIDGE LIBRARY COLLECTION

Books of enduring scholarly value

Physical Sciences

From ancient times, humans have tried to understand the workings of the world around them. The roots of modern physical science go back to the very earliest mechanical devices such as levers and rollers, the mixing of paints and dyes, and the importance of the heavenly bodies in early religious observance and navigation. The physical sciences as we know them today began to emerge as independent academic subjects during the early modern period, in the work of Newton and other 'natural philosophers', and numerous sub-disciplines developed during the centuries that followed. This part of the Cambridge Library Collection is devoted to landmark publications in this area which will be of interest to historians of science concerned with individual scientists, particular discoveries, and advances in scientific method, or with the establishment and development of scientific institutions around the world.

Star-Land

Sir Robert Stawell Ball's *Star-Land* of 1889 is based on some of his Christmas Lectures at the Royal Institution during his time as royal astronomer of Ireland, a post he held from 1874 to 1892. These lectures were aimed at a young audience in order to introduce them to the subject, and fire their interest in the wonders of the universe. This volume includes lectures on the sun, the moon, the inner and giant planets, comets and shooting stars, and stars. It also contains a chapter on the observation and naming of stars. Ball was a renowned public lecturer, with commissions across Britain, Ireland and the United States, where his anecdotal and conversational style won him much popularity. The author of several frequently reprinted science books, he was knighted in 1886 and in 1892 became Lowendean professor of astronomy at Cambridge and the director of the university observatory.

Cambridge University Press has long been a pioneer in the reissuing of out-of-print titles from its own backlist, producing digital reprints of books that are still sought after by scholars and students but could not be reprinted economically using traditional technology. The Cambridge Library Collection extends this activity to a wider range of books which are still of importance to researchers and professionals, either for the source material they contain, or as landmarks in the history of their academic discipline.

Drawing from the world-renowned collections in the Cambridge University Library, and guided by the advice of experts in each subject area, Cambridge University Press is using state-of-the-art scanning machines in its own Printing House to capture the content of each book selected for inclusion. The files are processed to give a consistently clear, crisp image, and the books finished to the high quality standard for which the Press is recognised around the world. The latest print-on-demand technology ensures that the books will remain available indefinitely, and that orders for single or multiple copies can quickly be supplied.

The Cambridge Library Collection will bring back to life books of enduring scholarly value (including out-of-copyright works originally issued by other publishers) across a wide range of disciplines in the humanities and social sciences and in science and technology.

Star-Land

Being Talks with Young People about the Wonders of the Heavens

Robert Stawell Ball

CAMBRIDGE UNIVERSITY PRESS

Cambridge, New York, Melbourne, Madrid, Cape Town, Singapore,
São Paolo, Delhi, Dubai, Tokyo

Published in the United States of America by Cambridge University Press, New York

www.cambridge.org
Information on this title: www.cambridge.org/9781108014175

This edition first published 1889
This digitally printed version 2010

ISBN 978-1-108-01417-5 Paperback

STAR-LAND.

By the same Author.

THE STORY OF THE HEAVENS.

WITH

Chromo Plates and Numerous Illustrations.

APPLIED MECHANICS.

Illustrated.

CASSELL & COMPANY, LIMITED: *London, Paris, New York & Melbourne.*

A JUVENILE LECTURE AT THE ROYAL INSTITUTION.

STAR-LAND.

BEING TALKS WITH YOUNG PEOPLE ABOUT THE WONDERS OF THE HEAVENS.

BY

SIR ROBERT STAWELL BALL, F.R.S.,

ROYAL ASTRONOMER OF IRELAND;

Author of "The Story of the Heavens," &c.

Illustrated.

CASSELL & COMPANY, LIMITED:

LONDON, PARIS, NEW YORK & MELBOURNE.

1889.

LA BELLE
SAUVAGE

To

THOSE YOUNG FRIENDS

WHO HAVE ATTENDED MY CHRISTMAS LECTURES

This little Book

IS DEDICATED.

PREFACE.

It has long been the custom at the Royal Institution of Great Britain to provide each Christmastide a course of Lectures specially addressed to a juvenile audience.

On two occasions, namely, in 1881 and in 1887, the Managers entrusted this honourable duty to me. The second course was in the main a repetition of the first; and on my notes and recollections of both the present little volume has been founded.

I am indebted to my friends Rev. MAXWELL CLOSE, Mr. ARTHUR RAMBAUT, and Dr. JOHN TODHUNTER for their kindness in reading the proofs.

ROBERT S. BALL.

OBSERVATORY,
Co. DUBLIN,
22nd Oct., 1889.

PREFACE

CONTENTS.

LECTURE I.

THE SUN.

LECTURE II.

THE MOON.

LECTURE III.

THE INNER PLANETS.

LECTURE IV.

THE GIANT PLANETS.

LECTURE V.

COMETS AND SHOOTING STARS.

LECTURE VI.

STARS.

CONCLUDING CHAPTER.

STAR-LAND.

LECTURE I.

THE SUN.

The Heat and Brightness of the Sun—Further Benefits that we receive from the Sun—The Distance of the Sun—How Astronomers Measure the Distances of the Heavenly Bodies—The Apparent Smallness of Distant Objects—The Shape and Size of the Sun—The Spots on the Sun—Appearances seen during a Total Eclipse of the Sun—Night and Day—The Daily Rotation of the Earth—The Annual Motion of the Earth round the Sun—The Changes of the Seasons—Sunshine at the North Pole.

THE HEAT AND BRIGHTNESS OF THE SUN.

We can all feel that the sun is very hot, and we know that it is very big and a long way off. Let us first talk about the heat from the sun. On a cold day it is pleasant to go into a room with a good fire, and everybody knows that the nearer we go to the fire, the more strongly we feel the heat. The boy who is at the far end of the room may be shivering with cold, while those close to the fire are as hot as they find to be pleasant. If we could draw nearer to the sun than we are at present, we should find ourselves to be warmer than we are here. Indeed, if we went close enough, the temperature would rise so much that we could not endure it; we should be roasted. On the other hand, we should certainly be frozen to death if we were transported much further away from the sun than we are now. We are able to live comfortably, because our bodies are just arranged to suit the warmth at that distance from the sun at which the earth is actually placed.

Suppose you were able to endure any degree of heat, and that you had some way of setting out on a voyage to the sun. Take with you a wax candle, a leaden bullet, a penny, a poker, and a flint. Soon after you have started you find the warmth from the sun increasing, and the candle begins to get soft and melt away. Still, on you go, and you notice that the leaden bullet gets hotter and hotter, until it becomes too hot to touch, and upwards the temperature continues to move, until at last the lead has melted, as the wax had previously done. However, you are still a very long way from the sun, and you have the penny, the poker, and the flint remaining. As you approach closer to the luminary the heat is ever increasing, and at last you notice that the penny is beginning to get red-hot; go still nearer, and it melts away, and follows the example of the bullet and the candle. If you still press onwards, you find that the iron poker, which was red-hot when the penny melted, begins to get brighter and brighter, till at last it is brilliantly white, and becomes so dazzling that you can hardly bear to look at it; then melting commences, and the poker is changed into liquid like the penny, the lead, and the wax. Yet a little nearer you may carry the flint, which is now glowing with the same fervour which fused the poker, but at last the flint too will have melted.

You will ask, how do we learn all this? for as nobody could ever make such a journey, how can we feel certain that the sun is so excessively hot? I know that what I say is true for various reasons, but I will only mention one, which is derived from an experiment with the burning-glass, that most boys have often tried.

We may use one of those large lenses that are intended for magnifying photographs. But almost any kind of lens will do, except it be too flat, as those in spectacles generally

Fig. 1.—How to use the Burning-glass.

are. On a fine sunny day in summer, you turn the burning-glass to the sun, and by holding a piece of paper at the proper distance a bright spot will be obtained (Fig. 1). At that spot there is intense heat, by which a match can be lighted, gunpowder exploded,. or the paper itself kindled. For the broad lens collects together the rays from the sun that fall upon it, and concentrates them on one point, which consequently becomes so hot and so bright. If we merely used a flat piece of glass the sunbeams would go straight through; they would not be gathered together, and

Fig. 2.—The Noonday Gun.

they would not be strong enough to burn. But the faces of the lens are specially curved, so that they have the effect of bending-in all the rays of light and heat, in such a way that they are all directed into one point, which we call the *focus.* When a great number of rays are thus collected on the same spot, each of them contributes a little warmth. And just as a penny a day put into a money-box makes more than thirty shillings by the end of the year, so the total effect of the several little heatings which the paper gets by the different rays suffices to set it on fire.

Some ingenious person has turned this principle to an

Fig. 3.—A Tell-tale for the Sun.

odd use, by arranging a burning-glass over a cannon in such a way, that just when noon arrived, the spot of light should reach the touch-hole of the cannon and fire it off. Thus the sun itself is made to announce the middle of the day. (Fig. 2.)

Another application of the burning-glass is to obtain a record of the number of hours of sunshine in each day. You will understand the apparatus from Fig. 3; the lens is here replaced by a glass globe. As the sun moves over the sky, the bright spot of light also moves, and will therefore burn its track on a sheet of paper marked with lines corresponding to the hours. When the sun is hidden by clouds the burning ceases, so by preserving each day the piece of paper, we have an unerring tell-tale, which shows us during what hours the sun was shining brightly, and the hours during which he was hidden. You see, the burning-glass is not merely a toy, it is here made useful in helping us to learn something about the weather.

Another experiment with the burning-glass will teach us something. Take a candle, and from its flame you can get a bright point at the focus. It may fall upon your hand, but you can hardly feel it, and you will readily believe that the focus is not nearly so hot as the candle. Even when a burning-glass is held in front of a bright fire, there is but little heat in the focus. By using a lens to condense the beams from an electric lamp, Professor Tyndall has shown us that we can light a piece of paper, and produce many other effects. But, nevertheless, it is not nearly so hot at the focus as in the actual arc, for you might move your finger through the focus without much inconvenience, but I would not recommend you to trust your finger between the poles of the electric light itself. It will always be found that the temperature obtained at the focus of a burning-glass is less than that obtainable at the source of heat itself. This must be true when we turn a burning-glass to the sun, and hence we know that the sun must be hotter than any heat which can be obtained by the biggest burning-glass on the brightest of summer days. But burning-glasses a yard wide have been made, and astonishing heat effects have been produced. Steel has thus been melted by the sunbeams, and so have other substances which even our greatest furnaces cannot fuse. Therefore the sun must itself be hotter than the temperature of molten steel; hotter, probably, than any temperature we can produce on earth.

I have tried to prove to you that the sun is very hot; but it would be well to see what arguments might be produced on the other side. Indeed, it is by considering objections that we often learn. So I shall tell you of a difficulty that was once raised when I was endeavouring to explain the heat of the sun to an intelligent man. "I am sure," said my friend, "that you must be quite wrong. You said that the nearer you got to the sun the hotter it would be; but I know this to be a mistake. When tourists go to Switzerland, they sometimes climb very high mountains. But on the top of a mountain you are, of course, nearer the sun than you were below; and so, if the sun were really hot you should have found it much warmer on the top of the mountain than at its base. But every one knows that there is abundant ice and snow on lofty Alpine summits, while down below in the valleys there is at the same time excessively warm weather. Does it not therefore seem that the nearer we go to the sun the colder it is, and the further we are from the sun the warmer it is?"

This is indeed a peculiar difficulty. The coldness of the mountain tops depends upon the fact that while there is something else besides the sun which contributes to keeping us warm, this something is more or less deficient at great heights. You know that we live by breathing air, and we find the air wherever we go, over land and sea, all round the earth. Those who ascend in a balloon are borne upwards by the air, and thus we can show that air extends for miles and miles over our heads, though it becomes lighter and thinner the loftier the elevation.

We not only utilise the air for breathing, but it is also of indispensable service to us in another way. It acts as a blanket to keep the earth warm ; indeed, we ought rather to describe the air as a pile of blankets one over the other. These air blankets enable the earth to preserve the heat received from the sunbeams by preventing it from escaping back again into space. Thus warmth is maintained, and our globe is rendered habitable. You see then that for our comfort we require not only the sun to give us the heat, but also the set of blankets to keep it when we have got it. If we threw off the blankets we should be uncomfortable, though the sun were as bright as before. It would be nearly as inconvenient to lose the useful blankets which alone enable us to keep in the heat after we receive it, as to be deprived of the sunbeams altogether. A man who goes to the top of a mountain at midday does get nearer to the sun, and, so far as this goes, he ought no doubt to feel warmer, but the gain is quite inappreciable. Even at the top of Mont Blanc the increase in heat due to the approach to the sun would be only one-ten-millionth part of the whole. This would be utterly inappreciable; no thermometer would even show it. On the other hand, by ascending to the top of the mountain he has got above the lower regions of the air ; he has not, it is true, reached even halfway to the upper surface—that is still a long way over his head—but the higher layers of the atmosphere are so very thin that they form most indifferent blankets. The alpine climber on the top of the mountain has thus thrown off the best portion of his blankets, and receives a chill ; while the gain of heat arising from his closer approach to the sun is imperceptible. Perhaps you will now be able to understand why eternal snow rests on the summits of the great mountains. They are chilled because they have not so many air blankets as the snug valleys beneath.

The brightness of the sun is among the most wonderful things in nature, and there are three points that I ask you to remember, and then indeed you will agree with Milton, that the sun is "with surpassing glory crowned." First think of the beauty and brilliancy of a lovely day in June. Then remember that all this flood of light comes from a single lamp at a most tremendous distance; and thirdly, recollect that the sun is not like a bull's-eye lantern, concentrating all his light *specially* for our benefit, but that he diffuses it equally around, and that we do not get on this earth the two-thousand-millionth part of what he gives out so plenteously! When we think of the brightness of day, which we love so much, and of the distance from which that light has come, notwithstanding that the sun dispenses with all assistance from condensing appliances, we begin to comprehend the sun's true magnificence.

FURTHER BENEFITS THAT WE RECEIVE FROM THE SUN.

I want to show you how great should be the extent of our gratitude to the sun. Of course, on a bright summer's day, when we are revelling in the genial warmth and enjoying the gladness of sunshine, it needs no words to convince us of the utility and of the beneficence of sunbeams. So we will not take midsummer. Let us take midwinter. Take this very Christmas season, when the days are short and cheerless, the nights are long and dark and cold. We might be tempted to think that the sun had well-nigh forgotten us. It is true he only seems to pay us very occasional visits, and between fogs and clouds we see but little of him; but, visible or invisible, the sun incessantly tends us, and provides for our welfare in ways that perhaps we do not always remember.

Let me give an illustration of what I mean. You will go back this dull and cold afternoon to the happy home where your Christmas holidays are being enjoyed. It will be quite dark ere you get there, for the sun in these wintry days sets so very early. You will gather around a cheerful fire. The curtains will be drawn, the lamps will be lighted, and the disagreeable weather outside will be forgotten in the pleasant warmth and light inside. Five o'clock has arrived, the pretty little table has been placed near mamma's chair, on it are the cups and saucers and the fancy teapot. Under the table is a little shelf, with some tempting cakes and a tender muffin. A welcome friend or two have joined the little group, and a delightful half-hour is sure to follow.

But you may say, "What have tea and muffins, lamps and fire-places to do with the sun? are they not all mere artificial devices, as far removed as possible from the sunbeams or the natural beauties which sunbeams create?" Well, not so far, perhaps, as you may think. Let us see.

Poke up the fire, and while it is throwing forth that delicious warmth, and charming but flickering light, we shall try to discover where that light and heat have come from. No doubt they have come from the coal, but then, whence came the coal? It came from the mine, where brave colliers hewed it out deep under the ground, and then it was hoisted to the surface by steam engines. Our inquiry must not stop here, for another question immediately arises, as to how came the coal into the earth? When we examine coal carefully, by using the microscope to see its structure, we find that it is not like a stone; it is entirely composed of trees and other plants, for different portions of leaves and stems can be recognised. Sometimes, indeed, the fossil trunks and roots of the great trees are found down in the coal-pit. It is quite plain, then, that these are only the remains of a vegetation which must once have been growing and flourishing, and we thus learn that coal must have been produced in the following manner:—

Once upon a time a great forest flourished. The sun shone down on this forest, and genial showers invigorated it, while insects and other creatures sported in its shades. It is true that the trees and plants were not like those we generally see. They were more like ferns and marestails and gigantic club-mosses. In the fulness of time they died, and fell, and decayed, and others sprang up to meet the like end. Thus it happens that, in course of ages, the remains of leaves, and fruits, and trunks accumulated over the soil. The forest was situated near the sea-shore, and then a remarkable change took place—the land began slowly to sink down. You need not be much surprised at this. Land has often been known to gradually change its level. In fact, a sinking process is slowly going on now in many places on the earth, while the land is rising in other localities. As the great forest gradually sank lower and lower, the sea-water began to flow over it, and the trees perished until, at last, deep water submerged the surface which had once been covered by a fine forest. At the bottom of this sea lay the decaying vegetation.

That which was the destruction of the growing forest, proved to be the means of preserving its remains, for then as now the rivers flowed into the sea, and the waters of the rivers, especially in times of flood, carried down with them much mud, which is held in suspension. Upon the floor of the ocean this mud was slowly deposited; and thus a coating of mud overlaid the remains of the forest. In the course of ages, the layers of mud grew thick and heavy, and hardened into a great flat rock, while the trunks and leaves underneath were squeezed together by the weight, and packed into a solid mass which became black, and in the course of time was transformed into coal.

After ages and ages had passed by, the bottom of the sea ceased to sink, and began slowly to rise. The sea over the newly made beds of stone became shallower, and at last the floor was raised until it emerged from the sea. But, of course, it would not be the original ground which formed the surface of the newly uncovered land. The sheets of deposit lay upon the forest; over the fresh surface life gradually spread, until man himself came to dwell there, while far beneath his feet were buried the remains of the ancient vegetation.

When we now dig down through the rocks we come upon the portions of trees and other plants which the lapse of time, and the influence of pressure, have turned from leaves and wood into our familiar coal.

That ancient forest grew because sunbeams abounded in those early times, and nourished a luxuriant vegetation. The heat and the light then expended so liberally by the sun, were seized by the leaves of flourishing plants, and were stored away in their stems and foliage. Thus it is that the ancient sunbeams have been preserved in our coal-beds for uncounted thousands of years. When we put a lump of coal on our fire this evening, and when it sends forth a grateful warmth and cheerful light, it but reproduces for our benefit some of that store of preserved sunbeams of which our earth holds so large a treasure. Thus, the sun has contributed very materially to our comfort, for it has provided the fire to keep us warm.

The sun has, however, ministered further to our tea party, for has it not produced the tea itself? The tea grew a long way off, most likely in China, where the plant was matured by the warmth of the sunbeams. From China the tea-chests were brought by a sailing vessel to London; the ship performed this long voyage by the use of her sails, which were blown by what we call wind, which is merely the passage of great volumes of air as they hurry from one part of the earth to another.

We may ask what makes the air move, for it will not rush about in this way unless there be considerable force to drive it. Here again we perceive the influence of the sun. Tracts of land are warmed by the sun. The air receives the heat from the land, and the warm air is buoyant and ascends while other cooler air continually flows in to supply its place. To do this it has, of course, to rush across the country, and thus wind is caused. All the winds on our earth are consequently due to the sun. You see, therefore, how greatly we are indebted to the sun for the enjoyment of our tea-table. Not only did the sun give us the coal and the tea, but it actually provided the means by which the tea was carried all the way from China to our own shores.

We can also trace the connection between the hot water and the sun. Of course the water has come immediately from the kettle, and that has been taken from the fire, and the fire was produced by sunbeams. Thus, we learn that it is the warmth of the sun that has made the water boil. The water itself came from the pipe that enters the house to give a supply of pure liquid from the waterworks. If you visit the waterworks, you will see great reservoirs. In some cases they will be filled from a river, sometimes the water is pumped from a deep well in the ground, sometimes it is the surface-water caught on a mountain side. Whatever be the immediate source of the supply, the real origin is to be sought, not in the earth beneath, but in the heavens above. All the water has come from the clouds. It is the clouds which sent down the rain, or sometimes the snow, or the hail, and it is this water from the clouds which fills our rivers. It is this water also which sinks deep into the earth and supplies our wells, so that from whatever apparent source the water seems to have come, it is indeed the clouds which have been the real benefactors. The water in the tea-pot to-night was, some time ago, in a cloud, floating far up in the sky.

We must look a little further and try to find from whence the clouds have come. It is certain that they are merely a form of steam or vapour of water, and as they are so continually sending rain down on the earth, there must be some means by which their supply shall be replenished. Here again our excellent friend the sun is to be found ever helping us secretly, if not helping us openly. He pours down his rich and warm beams on the great oceans, and the heat turns some of the water into vapour, which being lighter than the air ascends upwards for miles. There the vapour often passes into the form of clouds, and the winds waft these clouds to refresh the thirsty lands of the earth. Thus, you see, it is the sun which procures for us water from the great oceans which cover so much of our globe, and sends it on by the winds to supply our waterworks, and fill our tea-pots. Notice another little kindliness of our great benefactor. The water of the oceans is quite salt. But we could not make tea with salt water, so the sun, when lifting the vapour from the sea, most thoughtfully leaves all the salt behind, and thus provides us with the purest of sweet water.

That nice muffin was baked by the sun, and toasted by the sun, and made from wheat grown by the sun, but the wheat, of course, had to be ground into flour. If it was ground in a wind-mill, then the sun raised the wind which turned the mill. If the flour-mill were driven by steam, then the sun, long ago, provided the coal for the boiler. If the miller lived on a river, then he used a water-mill, but here again the sun did the work. The sun raised the water to the clouds, and after it had fallen in rain, and was on its way back to the sea, its descent was utilised to turn the water-wheel. The water derives its power to turn the mill from the fact that it is running down, but it could not run down unless it had first been raised up; and thus it is the sun which drives round the water-wheel. Nor can the baker dispense with the sun's aid even if he rejected wind-mills, or steam-mills, or water-mills, and determined to grind the corn himself with a pestle and mortar. Here, at least, it might be thought that it is a man's sinews and muscles that are doing the work, and so no doubt they are. But you are mistaken if you think the sun has not rendered indispensable aid. The sun has just as surely provided the power which moves the baker's arms as it has raised the wind which turned the wind-mill. The force exerted in grinding with the pestle has been derived from the food that the man has eaten; that food was grown by the sun, and the man received from the food the power it had derived from the sun's heat. So that, look at it any way you please, even for the grinding of the wheat to make the muffin for your tea party, you are wholly indebted to the sun.

It is the sun which has bleached the table-cloth to that snowy whiteness. The sun has given those bright colours which look so pretty in the girls' dresses. With how much significance can we say and feel that light is pleasant to the eye, and what prettier name than Little Sunbeam can we have for the darling child that makes our home so bright?

THE DISTANCE OF THE SUN.

The sun is a very long way off; it is not easy for you to imagine a distance so great, but you must make the attempt. This is the first measurement that we shall have to make on our way to that far-off country called Star-Land; but long as we shall find it to be, we shall afterwards have to span many distances very much longer. When you are out in the street, or taking a walk in the country, you know that this man is near, or that house is far, or that mountain is many miles away, because you have other objects between to help you to judge of the distances. You will see, for example, that there are many houses or farmyards, and you will notice hedges dividing different fields between you and the mountain. You also see that there are woods and parks, and perhaps stretches of moorland extending up the slopes. You have a general idea that the farmyards and fields are of considerable size, and that the woods or moors are wide and extensive; and putting these things together, you realise that the mountain must be miles away.

But when we look at the sun we have no aids conveniently placed to help us to judge of his distance. There are no intervening objects, and merely gazing at the sun helps us but little in obtaining any accurate knowledge. We must go to the astronomer and ask him to tell us how far he has found the sun to be, and then we must also beg from him some explanation of the method he has used in making his measurements.

It has been found that the sun is about ninety-three millions of miles from the earth; but sometimes it is a little further and sometimes it is a little nearer. Let us first try to count 93,000,000. The easiest way will be to get the clock to do this for us; and here is a sum that I would suggest for you to work out. How many times will the clock have to tick before it has made as many ticks as there are miles between the earth and the sun? Every minute the clock, of course, makes 60 ticks, and in 24 hours the total number will reach 86,400. By dividing this into 93,000,000 you will find that over 1,076 days, or nearly three years, will be required for the clock to perform the task.

We may consider the subject in another way, and find

how long an express train would take to go all the way from the earth to the sun. We shall suppose the speed of the train to be 40 miles an hour; and if the train ran for a whole day and a whole night without stopping, it would then accomplish 960 miles. In a year the distance travelled would reach 350,400 miles, and by dividing this into 93,000,000 we arrive at the conclusion that a train would have to travel at a pace of 40 miles an hour, not alone for days and for weeks and for years, but even for centuries. Indeed, not until 265 years had elapsed would the mighty journey have been ended. Even though King Charles I. had been present when the train began to move, the journey would not yet have been completed. No one who started in the train could expect to reach the end of the journey. That would not occur till the time of his great-great-grandchildren.

HOW ASTRONOMERS MEASURE THE DISTANCES OF THE HEAVENLY BODIES.

I shall so often have to speak of the distances of the celestial bodies that I may once for all explain how it is that we have been able to discover what these distances are. This is a very puzzling matter if we were to try and describe it fully, but the principle of the method is not at all difficult. Do you know why you have been provided with two eyes? It is generally said that one of the reasons is to aid you in estimating distances. You see this boy (Fig. 4) judges of the distance of his finger by the inclination of his two eyes when directed at it. In a similar way we judge of the distance of a heavenly body by making observations on it from two different stations.

I shall illustrate our method of measuring the actual distance of a body in the heavens by showing you how we can find the height of that large india-rubber ball which is hanging from the ceiling. Of course, I do not intend to have a measuring tape from the ball itself, because I want to ascertain its height on the same principle as that by which

Fig. 4.—Two eyes are better than one.

we measure the distance of the sun or of any other celestial body which we cannot reach. I will ask the aid of a boy and a girl, who will please stand one at each end of the lecture-table. The apparatus we shall want will be very simple: it will consist of two cards and a pair of scissors. The boy will kindly shape his card to such an angle that when he holds it to his eye one side of the angle shall point straight at the little girl, and the other side shall point straight at the ball, just as you see in the picture (Fig. 5). The girl will also please do the same with her card, so that along one side she just sees the little boy's face, while the other side points up

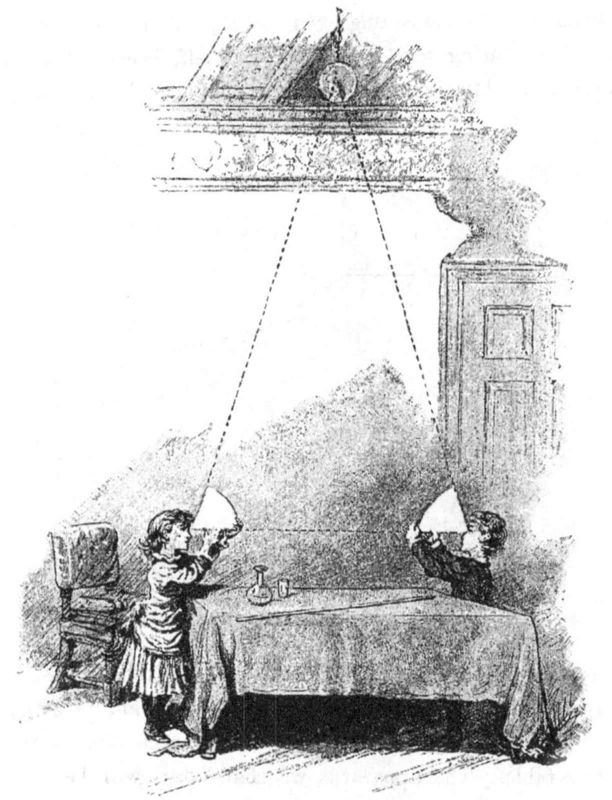

Fig. 5.—How we measured the height of the ball.

to the ball. It will be necessary to cut these angles properly. If the angle be too big, then when one side points to the boy's face, the other will be directed above the ball. If the angle on the card be too small, then one side will be directed below the ball, while the other is pointed to the boy. The whole accuracy of our little observations depends upon cutting the card angles properly. When they have been truly shaped it will be easy to find the distance of the ball. We first take a foot rule and measure the length of our table from the boy to the girl. That length is twelve feet, and to discover the distance of the ball we must make a drawing. We get a sheet of paper, and first rule a line twelve inches long. That will represent the length of the table, it being understood that each inch of the drawing is to correspond to a foot of the actual table. Let the end where the girl stood be marked B, and that of the boy, A, and now bring the cards and place them on the line just as shown in the figure. The card the girl has shaped is to be put so that the corner of it lies at B, and one edge along B A. Then the boy's card is to be so put that its corner is at A and one edge along A B. Next with a pencil we rule lines on the other edges of the cards, taking care that they are kept all the time in their proper positions. These two lines carried on will meet at C; and this must be the position of the ball on the scale of our little sketch. It only now remains to take the foot rule and measure on the drawing the length from A to C. I find it to be twenty inches, and I have so arranged it that the distance from B to C is the same.

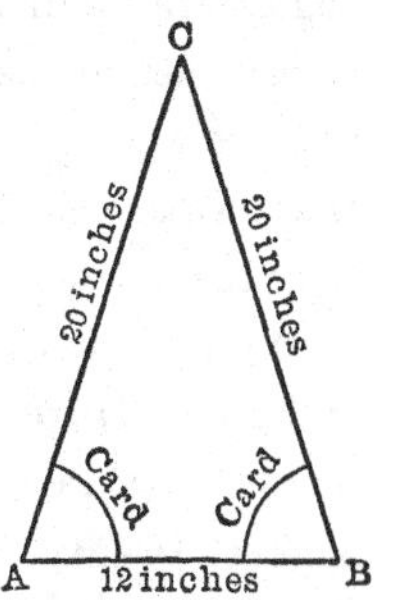

Fig. 6.—This is what we wanted the cards for.

I do not intend to trouble you much with Euclid in these

lectures, but as many of my young friends have learned the sixth book, I will just refer to the well-known proposition, which tells us that the lengths of the corresponding sides of two similar triangles are proportional. We have here two similar triangles. There is the big one with the boy at one corner, the girl at the other, and the ball overhead. Here is the small triangle which we have just drawn. These triangles are similar because they have got the same angles, and it was to ensure that they should have the same angles that we were so careful in shaping the cards. As these two triangles are similar, their sides must be proportional. We have agreed that the line A B, which is twelve inches long, is to represent the length of the table between the little boy and girl. Hence the distance, A C, must, on the same scale, be the interval between the ball and the boy at the end. This is twenty inches on the drawing, and therefore the actual distance from the end of the table to the ball is twenty feet.

Hence you see that without going up to the ball or having a twine from it, or in any other way making direct communication with it, we have been able to ascertain how far up in the air the ball is actually hung. This simple illustration explains the principle of the method by which astronomers are able to learn the distances of the different celestial bodies from the earth. You must think of the sun, the moon, and the stars, as globes supported in some manner over our heads, and we seek to discover their distances from measurements of angles made at the ends of a base-line.

Of course, astronomers must choose two stations which are far more widely separated than are those in our little experiment. In fact, the greater the interval between the two stations, the better. Astronomers require a much longer distance than from one side of this room to the other, or from one side of London to the other side. If it were merely a balloon at which we were looking, then, when one observer at one side of London and another at the opposite side shaped their cards carefully, we should be able to tell the height of the balloon very easily. But as the sun is so much further off than any balloon could ever be, we must separate the observers much more widely. Even the breadth of England would not be enough, so we have to make them separate more and more until they are as widely divided as it is possible for any two people on this earth to be. One astronomer goes to one side of the earth at A (Fig. 7), and the other to the opposite side at B, so that they can both see the sun. They are obliged to use a much more accurate way of measuring the angles than by cutting out cards with pairs of scissors; and as the astronomer at A is not able to see his friend at B, it becomes no easy matter to measure the angles accurately. However, we need not trouble about the difficulties; it will suffice that the angles are measured by the astronomers in some way, and then they make a little sketch just as we did, or they make a calculation which is equivalent to the sketch. The astronomers know the size of

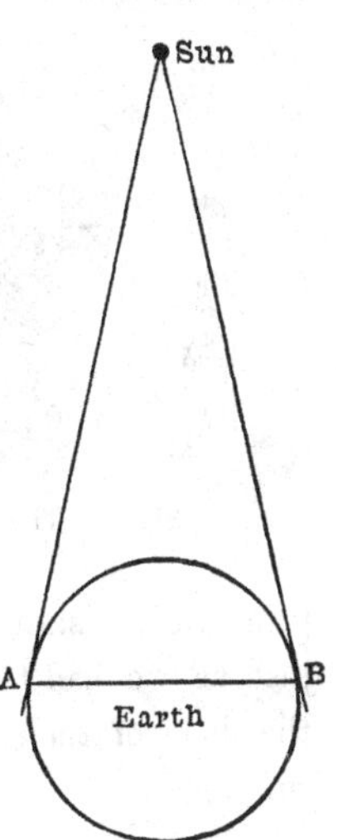

Fig. 7.—This would be our base-line when finding the Sun's distance.

the earth, and thus they know how many thousands of miles lie between them when making the observations. This distance means in their drawing what the length of the table did in ours. From each end of the line they set off an angle just as we did, and the astronomer must use the

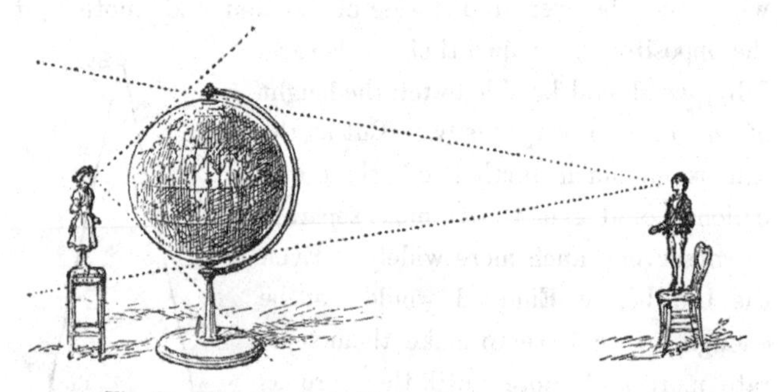

Fig. 8.—The globe looks bigger the nearer you are to it.

principle of similar triangles which he finds in Euclid, just as we had to do. At last, when they have measured the sides of their triangle, they obtain the distance of the sun.

THE APPARENT SMALLNESS OF DISTANT OBJECTS.

I ought here to explain a principle which those who are learning about the stars must always bear in mind. It asserts that the further a body is the smaller it looks. Perhaps this will be understood from the adjoining little sketch (Fig. 8). It represents a great globe, on which oceans and continents are shown, and you see a little boy and a little girl are looking at this globe. The girl stands quite close to it, and I have drawn two dotted lines from her eye, one to the top of the globe, and the other to the under surface. If she wants to examine the entire side of the globe which is towards her, she must first look along the upper dotted line, and then she must turn her glance downwards until she comes to the lower line, and having to turn her eyes thus up and down she will think the globe is very big, and she will be quite right. The boy is, as you see, on the other side of the globe, but I have put him much further off than the girl. I have also drawn two dotted lines from his eye to the globe, and it is plain that he will not have to turn his head much up

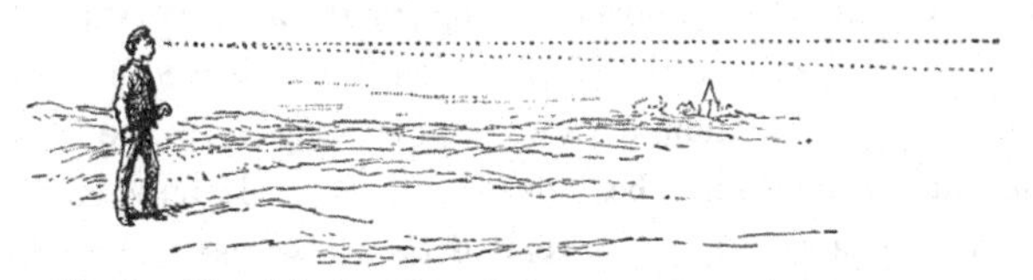

Fig. 9.—The globe is so far off that it lies beyond the picture. The dotted lines show how small it seems.

and down to see the whole globe. He can take it all in at a glance, and to him, therefore, the globe will appear to be comparatively small, because he is sufficiently far from it. The more distant he is, the smaller it will appear. You can easily imagine that, if the globe were far enough, the two lines that would include the whole would be like those shown (Fig. 9), in which the globe is so distant that it cannot be seen in the picture. The apparent size which is really measured by the angle between these two lines would be ever smaller and smaller the greater was the distance. Now you can understand how it is that the further an object is the smaller it looks; indeed, when sufficiently far, the object ceases to be visible altogether.

I could give many illustrations of the diminution of size by distance, and so, doubtless, could you. Every boy knows that his kite looks smaller and smaller the greater the length of string that he lets out. I have seen in the West of Ireland a bird that seems like a little speck high up near the clouds, but from its flight and other circumstances I knew that the speck was not really a little bird. It was, indeed, a great eagle, which had been dwarfed by the elevation to which he had soared.

It is in astronomy that we have the best illustrations of this principle. Enormous objects seem to be small because they are so very far off. You must therefore always remember that although an object may appear to be small, this appearance may be only a delusion. It may be that the object is very big, but very distant. In astronomy, this is almost always the case, for there is so much room above us, around us, on all sides in space. Look up at the ceiling. It certainly does not bound space, for there is another side to it; and then there is the roof of the house. But the roof is not a boundary, for, of course, there is the air above it, and then, higher up still, there are the clouds, and so we can carry our imagination on and on through and beyond the air up to where the stars are, and still on and on. There can be no bounds to our thoughts, for space has no ending. And as there is unlimited room, the celestial bodies take advantage of it, and are, generally speaking, at such gigantic distances that, no matter how small they may appear, that smallness is merely a deception.

Let us try to illustrate in another way the exceeding remoteness of the sun. So please imagine that you were on the sun, and that you took a view of our earth from that distance. To find out what we must expect to see, let us think of a balloon voyage. If you were to go up in a balloon, you would at first only see the houses, or objects immediately about you, but as you rose the view would become wider and wider. You would see that London was surrounded by the country, and then, as you still soared up and up, the sea would become visible, and you would be able to trace out the coasts, east and west and south. If, in some way, you could soar higher than any balloon could carry you, the whole of the British Islands would presently lie spread like a map beneath. Still on and on, and then the continent of Europe would be gradually opened out, until the great oceans, and even other continents, would at last be glimpsed, and then you would perceive that our whole earth was indeed a globe. The higher you went, the less distinctly would you be able to see the details on the surface. At last the outlines of the continents and oceans would fade, and you would begin to lose any perception of the shape of the earth itself. Long ere you had reached the distance of the sun, the earth would look merely as the planet Venus now does to us. Indeed, the sun is so far off, that it is instructive to consider how small our earth would seem when viewed from that distance. Think of that very familiar little globe, a tennis-ball, which is two and three-quarter inches in diameter. But suppose a tennis-ball were at the opposite side of the street, or still further away; suppose, for example, that it were half a mile away, what could you expect to see of it? and yet the earth, as seen from the sun, would appear to be no larger than a tennis-ball would look when viewed from a distance of half a mile.

THE SHAPE AND SIZE OF THE SUN.

We have spoken of the heat of the sun, how hot he is; of the distance of the sun, how far he is; and now we must say a little about the size of the sun; and also about his shape. It is plain that the sun is round, that it has the shape of a ball. We are sure of this because, though a plate is circular, yet, if it were placed so that we only saw it edgeways from a distance, it would not appear to be circular. The sun is always turning round, and as it always seems to be a circle, we are therefore certain that the true shape of the sun must be globular, and not merely circular like a flat plate.

In the middle of the day, when the sun is high in the heavens, it is impossible for us to form a notion of the size of the sun. People will form very different estimates as to his apparent bigness. Some will say he looks as large as a dinner plate, but such statements are meaningless, unless we say where the plate is to be held. If it be near the eye, of course the plate may hide the sun, and, for that matter, everything else also. If the plate be about a hundred feet away, then it might just about hide the sun. If the plate were more than a hundred feet distant, then it could not hide the sun entirely, and the further the plate, the smaller it would seem.

No means of estimating the sun's size are available when he is high in the heaven. But when he is rising or setting, we see that he passes behind trees or mountains, so that there are intervening objects with which we can compare him; then we have actual proof that the sun must be a very large body indeed.

I give here a picture, by Marcus Codde, taken from a

French journal, *l'Astronomie,* which gives a charming illustration of a sunset at Marseilles (Fig. 10). If you wish to see that the sun is bigger than a mountain, you may go to the top

Fig. 10.—A Sunset viewed from Marseilles. (Marcus Codde.)

of Notre Dame de la Garde, but you must choose either the 10th of February or the 31st of October for your visit, because it is only on the evenings of those days that the sun sets in the right position.

On both these evenings the sun sinks directly behind

Mount Carigou in the Pyrenees; this mountain is a long way from Marseilles—no less, indeed, than one hundred and fifty-eight miles. But the mountain is so lofty, that when the sky is clear, the summit can be distinctly seen upon the sun as a background, in the way shown in the picture. This must be a very pretty sight, and it teaches us an important lesson. The sun is further away than the mountain, and yet you see the sun on both sides of the mountain, and above it. Here then we learn without any calculations, that the sun must be bigger than the upper part of a great mountain in the Pyrenees.

When we calculate the size of the sun from the measurements made by astronomers, we discover that it is much bigger than Mount Carigou; we see that even the entire range of the Pyrenees, the whole of Europe, and even our whole globe, are insignificant by comparison.

There is a football on the table, shown in Fig. 11. We shall suppose it to represent the sun; we shall now choose something else to represent the earth. We must, however, exhibit the proportions accurately. A tennis-ball will not do; it is far too large. The fact is, the width of the earth is less than the one-hundredth part of the width of the sun. The tennis-ball is, however, only a quarter the width of the football, so we must choose something a good deal smaller. I try with a marble, even with the smallest marble I could find, but when I measure it, I find that one hundred such marbles, placed side by side, would be far longer than the width of the football; I must therefore look for something still smaller. A grain of small-sized shot will give the right size for the model of our earth. About one hundred of these grains placed side by side will extend to a length equal to the width of the football. Now you will be able to form some conception of how enormous the sun really is. Think of this earth, how big we find it when we begin to travel. What a tremendous voyage we have to take to get to New Zealand, and even then we have only got half-way round the globe. Then think that the sun is in the same

Fig. 11.—How we compare the Earth and the Sun.

proportion bigger than the earth as that football is bigger than that grain of shot. If a million of such grains of shot were melted and cast into one globe, it would not be as large as that football. If a million globes, as large as our earth, could be united together, no doubt a vast globe would be produced, but it would not be as large as the sun. Think of a single house, with three or four people living in it, and then think of this mighty London, with its millions of inhabitants. The house will represent our earth, while great London represents the sun!

THE SPOTS ON THE SUN.

I have shown you that the sun is intensely hot, and a

Fig. 12.—A Party of Young Astronomers looking at the Sun.

very long way off, and enormously big. And now we have to see what the actual nature of the sun appears to be.

If you get a piece of very dark glass, or if you smoke a piece of glass over a candle, then you can look directly at the sun with comfort. A nicer plan is to prick a pinhole in a card, through which you can look at the sun without any inconvenience. Generally speaking, a view of the sun in this way will show you only an uniformly bright surface. To study the face of our great luminary carefully, you must use the aid which the telescope gives to the astronomer.

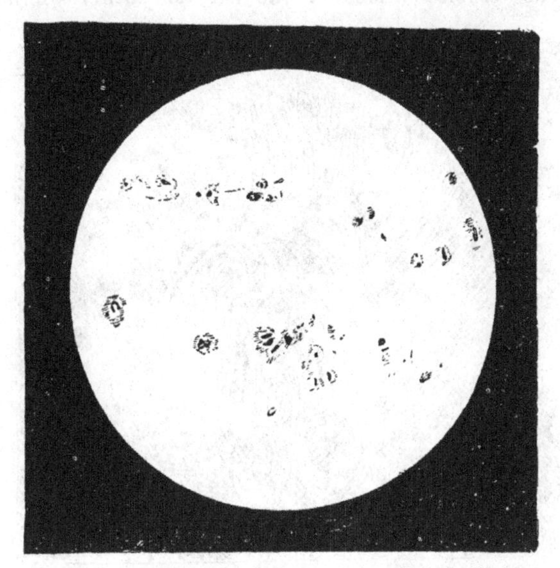

Fig. 13.—This is what the Sun sometimes looks like.

The correct way of doing this is shown in the opposite picture, Fig. 12. A small telescope, fixed on a stand, is pointed to the sun, and on a screen the sun draws its own picture. This may be examined without any inconvenience, or without the necessity for any protection to the eye, and a number of young astronomers can all view the sun at the same moment. On such a picture you will generally see the brilliant surface marked with dark spots, which are sometimes as numerous as shown in Fig. 13. When one of the sun-spots has been examined more closely, and with a much more powerful telescope, the wonderful structure represented in Fig. 14 has been seen.

The visible surface of the sun is entirely formed

Fig. 14.—A Sun-spot. (After Langley.)

of intensely heated vapours. We might almost say that the spots are holes, by which we can look through the brilliant surface to the interior and darker parts. Sometimes the spots close up, and fresh ones will open elsewhere. Sometimes the whole surface is mottled over in a remarkable way. I give here a picture which was taken from Mr. Nasmyth's beautiful drawing, in which he shows how the sun sometimes assumes the appearance which has been likened to willow-leaves (Fig. 15).

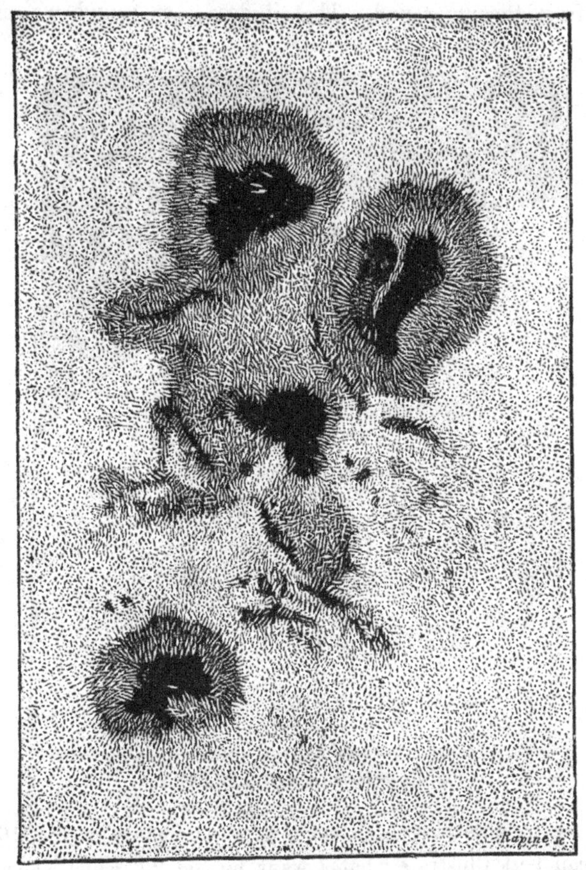

Fig. 15.—Nasmyth's Drawing of the Willow-leaved Structure of the Sun.

The spots often last long enough to demonstrate a remarkable fact. We must remember that the sun is a great globe, and that it is poised freely in space. There is nothing to hold it up, and there is nothing to prevent it from turning round. That it does turn round, we can

Fig. 16.—Spot nearing the Sun's Edge.

prove by observing such spots as those shown in Fig. 16. I can best illustrate what I want by Fig. 17, which shows six imaginary pictures. The first represents the sun on the 1st day of the month; the next shows it five days later, on the 6th; another view is five days later still, on the 11th;

and so on until the last picture, which corresponds to the 26th. You see, on the first day there is a spot near the left edge; by the 6th, this spot is near the middle; by the 11th, it is near the right edge; then you do not see it at all on the 16th, or on the 21st; but on the 26th it is back in the same place from which it started. We find other spots to have a similar history. They appear to move across the

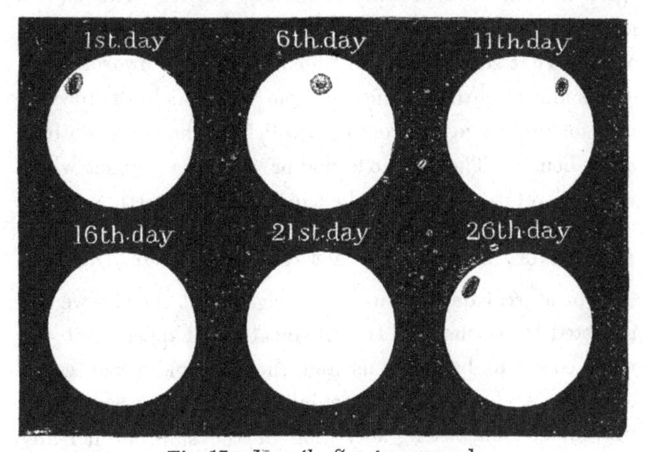

Fig. 17.—How the Sun turns round.

face, and then to return after twenty-five days to the same place where they were originally noticed. This appearance can be explained very simply by a little model. Take a white globe on which a black spot has been painted. I turn the globe round slowly with the handle, and the black spot just goes through the changes that we have seen. We start with the spot near the left, it moves across the face, and then passes to invisibility by moving behind the globe until it reappears again, after having moved round the back. As the same may be observed with every spot which lasts long enough, we learn that the changes in the places must be produced by the turning round of the sun. Here you see is the way in which an astronomical discovery is made. We first observe the fact that the spots do always appear to move. Then we try to account for this, and we find a very simple explanation, by supposing that the whole sun, spots and all, turns steadily round and round. This motion is always going on, and there are some curious consequences of it. That side of the sun which is turned towards us to-day is almost entirely different from that which was towards us a fortnight ago, or from that which we shall see in a fortnight hence. There is no actual or visible axis about which the sun rotates. In this the sun is like the earth.

APPEARANCES SEEN DURING A TOTAL ECLIPSE OF THE SUN.

For a great deal of our knowledge about the sun we are indebted to the moon. It will sometimes happen that the moon comes in between us and the sun, and produces an eclipse. At first you might think that an eclipse would only prevent us from seeing anything of the sun, but it really reveals to us beauties of which we should otherwise be ignorant. The sun has curious appendages which are quite hidden under ordinary circumstances. In the full light of day the dazzling splendour of the sun obliterates and renders invisible objects which shine with light which is comparatively feeble. It fortunately happens that the moon is just large enough to intercept the whole of the direct light from the sun, or rather, I should say, from the central parts of the sun. Surrounding that central and more familiar part from which the brilliancy is chiefly derived is a sort of fringe of delicate and beautiful objects which are self-luminous no doubt, but with a light so feeble that when the full blaze of sunlight is about they are invisible. When, however, the moon so kindly stops all the stronger beams, then these faint objects spring into visibility, and we have the exquisite spectacle of a total eclipse. The objects that I desire to mention particularly are the corona and the prominences.

A pretty picture of the total eclipse of the sun which occurred on 6th May, 1883, is here shown (Fig. 18.) It is taken from a drawing made by M. Trouvelot, who was sent out with a French observing party. They went a very long way to see an eclipse, but what they saw recompensed them for all their trouble. The track along which the phenomenon could be best seen lay in the Pacific Ocean, and an island was chosen which was so situated that the sun should be high in the heavens at the important moment, and also that the duration while the total eclipse lasted should be as long as possible. They accordingly went to Caroline Island, and all this journey to the other side of the earth was taken to witness a phenomenon that only lasted five minutes and twenty-three seconds. The eclipse lasted over the earth for more than five hours. During the greater part of that time the sun was only partially cut off by the interposition of the moon. It was only at the middle of the period that the moon was centrally on the sun. Short though these precious minutes were, they were long enough to enable good work to be done. Careful preparations had been made so that not a moment should be thrown away. Each member of the party had his special duty allotted to him, and this had been rehearsed so carefully beforehand that when the long-expected moment of "totality" arrived there was neither haste nor confusion; every-

Fig. 18.—Total Eclipse of the Sun, May 6, 1883. (Drawn by Trouvelot.)

one carefully went through his part of the programme. M. Trouvelot, for instance, occupied himself for two minutes and a few seconds in making the sketch that we now show. No doubt an accomplished astronomical artist like M. Trouvelot would gladly have had longer time for his sketch of so unique a sight, but brevity was imperative. He had taken pains previously to give accuracy to his work by arranging a little circle in the telescope with marks around it, so as to be able to show the positions of the different parts of the object. He had already had experience of similar eclipses, so that he was prepared at once to note what ought to be noted, and the picture we have shown is the result. This was completed within less than half of the duration of totality, and the artist had still three minutes left to devote to another and quite different part of the work, which does not concern us at present.

I want you particularly to look at these long branches or projections which we see surrounding the sun when totally eclipsed. They shine with a pearly light, and, in fact, it is stated that even during the gloomiest portion of the time there was still as much illumination as on a bright moonlight night. All that light was dispensed from this glorious halo round the sun which astronomers call the "corona." We do not under ordinary circumstances see a trace of this object. Even during a partial eclipse of the sun it is not visible, but directly the moon quite covers the sun, so as to cut off all the direct light, then the corona springs into visibility. It is always there, no doubt, though we cannot see it.

The corona is constantly changing in appearance. It seems to have little more permanence than the fleeting clouds in our air. It is evidently composed of some very unsubstantial material.

The other appendages to the sun which can be seen during an eclipse are the objects which we call "prominences." They are of a ruddy colour, and seem to be great flames, which leap upwards from the glowing surface of the sun below. Though the existence of the prominences was first discovered by their presence during eclipses, it fortunately happens that we are no longer wholly dependent on eclipses for the purpose of seeing them. It is true that we may look at the sun with even the biggest telescope in the ordinary way, and still not be able to perceive anything of the prominences. We require the aid of a special appliance with a glass prism to show them. But I am not now going to describe this ingenious contrivance. I am only going to speak of the results which have been obtained by its means. We shall here again avail ourselves of the experience of M. Trouvelot for a picture of two of these wonderful appendages.

The view (Fig. 19) shows the ordinary aspect of the sun diversified with the groups of dark spots. The fringe around the margin is of some ruddy material, forming the base of the flames which rise from the glowing surface. No doubt these flames are also often present on the face of the sun, but we cannot see them against the brilliant background. They are only perceptible when shown against the deep black sky behind. At two points of this margin, which happen curiously enough to be nearly opposite to each other, two colossal flames have burst forth. They extend to a vast distance, which is about one-third of the width of the sun. The vigour of these outbreaks may be estimated by the flickering which they exhibit; only, considering their size, we must allow them a little more time than is demanded for the movements of flames of ordinary dimensions. The great flame on the left was obviously declining in brilliancy when first seen. In a

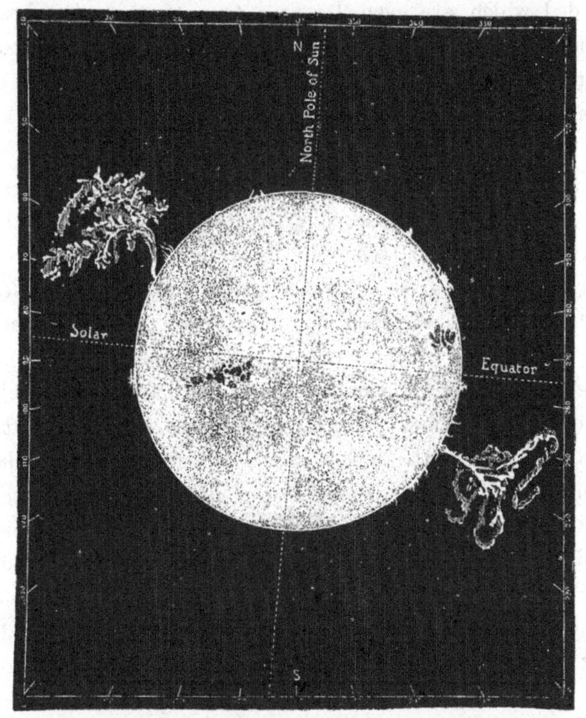

Fig. 19.—Solar Prominences. (Drawn by Trouvelot.)

quarter of an hour it had broken up into fragments, some of which were still to be seen floating in the sun's atmosphere. In ten minutes more the light of this flame had almost entirely vanished. Surely these are changes of extraordinary rapidity when we remember the size of this prominence. It is nearly 300,000 miles in height—that is to say, about thirty-seven times the width of our earth.

Great as are these prominences, others have been recorded which are even larger. One of them has been seen to rush up with a speed of 200,000 miles an hour—that is with more than two hundred times the pace of the swiftest of rifle-bullets.

NIGHT AND DAY.

There is a notable point of difference between the earth and the sun. The sun is bright, and the earth is dark. The sun gives light and heat, and the earth receives light and heat. We should be in utter darkness were it not for the sun; at least, all the light we should have, beyond the trivial lamps and candles, would come from the feeble twinkle of the stars. The moon would be no use to us, for the brightness of the moon is merely the reflection of the sunbeams. Were the sun gone, we could never again see the moon, and we should also miss from the sky a few other bodies, such as Jupiter and Venus, Mars and Saturn. But the stars would be the same as before, for they do not depend upon the sun for their light. We shall, indeed, afterwards see that each star is itself a sun.

Picture to yourself the earth as a target for sunbeams. These beams shoot down on one half of the globe, and give to it the brilliancy of day. The other half of the earth is at the same time in shadow, and there night reigns. The boundary between light and darkness is not quite sharply defined, for the pleasant twilight softens it a little, so that we pass gradually from day to night. Looking at the progress of the sun in the course of the day, we see that he rises far away in the east, then he gradually moves across the heavens past the south, and in the evening declines to the west, sets, and disappears. All through the night the sun is gradually moving round the opposite side of the earth under our feet, illuminating New Zealand and Japan and other remote countries, and then gradually working round to the east, where he starts afresh to give us a new day here.

Our ancestors many ages ago did not know that the earth was round. They thought it was a great flat plain, and that it extended endlessly in every direction. They were, however, much puzzled about the sun. They could see from the coasts of France and Spain or Britain that the sun gradually disappeared in the ocean : they thought that it actually took a plunge into the sea. This would certainly quench the glowing sun ; and some of the ancients used to think they heard the dreadful hissing noise when the great red-hot body dropped into the Atlantic. But there was here a difficulty. If the sun were to be chilled down every evening by dropping into the water hundreds of miles away to the west, how did it happen that early the next morning he came up as fresh and hot as ever, hundreds of miles away to the east? For this, indeed, it seemed hard to account. Some said that we had an entirely new sun every day. The gods started the sun far off in the east, and after having run its course it perished in the west. All the night the gods were busy preparing a new sun to be used on the succeeding day. But this was thought to be such a waste of good suns that a more economical theory was afterwards proposed. The ancients believed the continents of the earth, so far as they knew them, were surrounded by a limitless ocean. At the north, there were high mountains, and ice and snow, which they thought prevented access to the ocean from civilised regions. Vulcan was the presiding deity who navigated those wastes of waters, and to him was entrusted the responsible duty of saving the sun from extinction. He had a great barge ready, so that when the sun was just dropping into the ocean at sunset he caught it, and during all the night he paddled with his glorious cargo round by the north. The glow of the sun during the voyage could even be sometimes traced in summer over the great highlands to the north. This, at all events, was their way of accounting for the long mid-summer twilight. After a tedious night's voyage Vulcan got round to the east in good time for sunrise. Then he shot the sun up with such terrific force that it would go across the whole sky, and then the industrious deity paddled back with all his might by the way he had come, so as to be ready to catch the sun in the evening, and thus repeat his never-ending task.

THE DAILY ROTATION OF THE EARTH.

Vulcan and his boat seemed a pretty way of accounting for the sun's apparent motion. The only drawback was that it was all work and no play for poor Vulcan. There were also a few other difficulties. Captains of ships told us that they had sailed out on the great sea, and that so far from finding that the ocean extended on and on for ever, the water seemed to bend round, so that, in fact, after sailing far enough in the same direction, they found that they would be brought back again to the place from which they started. They also knew a little about the north. They told us that there was no sea there for Vulcan to navigate. It also appeared that ships had been voyaging all over the globe night and day in every direction, and that no captain had ever seen the sun coming down to the sea, and still less had he ever met with Vulcan in the course of his incessant voyages. Thus it was discovered that the earth could not be a never-ending flat, but that it must be a globe, poised freely in space without any attachment to hold it up. It was thought that the change from day to night might be accounted for by supposing that the sun actually went round the earth through the space underneath our feet. This is, indeed, what it seems to do. But there was a great difficulty about this explanation, which began to be perceived when the size and distance of the sun were considered. It required the sun to possess an alarming activity. He would actually have to rush round a vast circle, and complete the voyage once every day.

A much more simple explanation was available. It was shown that the sun did not revolve around the earth once every day, but that the earth itself turned round in such a way as to produce the changes from day to night. We may illustrate the case by this figure (Fig. 20). The small globe is the earth, which I can turn by the handle. The lamp will represent the sun, and as at present shown, the side of the earth, on which England lies, E, is towards the lamp and in full day. On the opposite side of the globe are other countries such as New Zealand, and there it is dark. You see that by simply turning the handle I can move England gradually round so that it passes into the dark side, and then night falls over the country. At the same time New Zealand is turned round to enjoy the smiles of day. This is a very simple method of accounting for the succession of day and night, and it is also the true method. We have already seen that the sun turns round, and now we find that the earth also turns, but the little body, the earth, goes much the faster, for it makes twenty-five turns while the sun goes round once.

Fig. 20.—How we illustrate the Changes between Day and Night.

Our earth is at this moment spinning round at a speed so great that London moves many hundreds of miles every hour. A town near the equator would gallop round at a pace of more than a thousand miles an hour—quicker, in fact, than a rifle-bullet. Don't you think that we ought to perceive that we are being whirled about in this terrific fashion? We know that when we are flying along in a railway train, we feel the jolting and we hear the noise, and we feel the blast of air if we put our heads out of window, and we see the trees as they rush past. All these things tell us that we are in rapid motion. But suppose these sensations were absent. Imagine a line so perfectly laid that no jolts should be perceptible, and that no racket can be heard; draw down the blinds so that nothing could be seen, how then are we to know of the motion? Indeed, your grandfathers used to be able to enjoy such a tranquil locomotion. I remember seeing in my childhood the fly-boats, as they were called, on the Royal Canal, wherein passengers were conveyed from Dublin to the West of Ireland, before the railway was made. The fly-boat was a sort of Noah's ark in appearance, drawn by a horse cantering along the towing path. In the cabin of such a vessel, where there was not the slightest motion of rolling or pitching—nothing but noiseless gliding along the canal—no one would be conscious of motion, so long as he did not look through the cabin windows. No one was ever sea-sick in a fly-boat; it was the perfection of travelling for those who loved ease and quiet.

The motion of the earth round its axis is, so far, like that of the fly-boat. It is so absolutely smooth that we do not feel anything, and we only become conscious of it by looking at outside objects. These are the sun, or the moon, or the stars. We see these bodies apparently going through their unvarying rising and setting, just as in looking out from the fly-boat, the passengers in that quaint old conveyance could see the houses and trees as they passed.

Seeing is believing; and I should like here, in this very theatre, to show you that we are actually turning round;

and this I am enabled to do by the kindness of my friend Professor Dewar, whom you all know so well.

I am tempted to wish for the moment that I had Aladdin's lamp, for I would rub it, and when the great genie appeared, I would bid him take the Royal Institution, and all of us here, to a place which everybody has heard of, and nobody has seen, I mean the North Pole. It would be so easy to describe the experiment I am about to show you, there. It is not so easy here. But it will be sufficiently accurate for our purpose to suppose that we actually have made the voyage, and that this is the Pole at the centre of the lecture-table. The direction of the axis round which the earth is turning is a line pointing up straight to the ceiling. This lecture-table and all the rest of the theatre is going round. In about six hours it will have moved a quarter of the way, and in twenty-four hours it will have gone completely round. That is, at least, what would happen if we were actually at the Pole. As we are not there, for the Pole is still many miles away from the Royal Institution, I must slightly modify this statement, and say that the table here takes more than twenty-four hours to go round. And now I want some way of testing the table to show that it actually is turning round. There is no use in our merely looking at it, because we ourselves, and this whole building, and the whole of London are all turning together. What we want is something which does not turn. Here is a heavy leaden ball. It is fastened to the roof by a fine steel wire, and you see it swings to and fro with a deliberate and graceful motion. I want it to oscillate very steadily, so I tie it by a piece of thread to a support, and then I burn the thread, and the great ball begins to swing to and fro. It would continue to do so for an hour, or indeed for several hours, and it is a peculiarity of this motion that the vibration always remains

Fig. 21.—A Pendulum.

in the same direction. Even the rotation of the earth will not affect the plane of this great pendulum, so far at least as our experiment is concerned. Here then we have a method of testing my assertion about the turning round of this theatre. I mark a line on the table, directly underneath the motion of the ball to and fro. If we could wait for an hour or so, we would see that the motion of the ball seemed to have changed into a direction inclined to its original position, but it is really the table that has moved, for the direction of the ball's motion is unaltered. We cannot, however, wait so long, therefore I show you the ingenious method which Professor Dewar has devised. By a beam from the electric light he has succeeded in so magnifying the effect that even in a single minute it is quite obvious that the whole of this room is distinctly turning round, with respect to the oscillations of the pendulum. This celebrated experiment proves by actual inspection that the earth must be rotating. By measuring the motion we can even calculate the length of the day, though I do not say it would be an accurate method of doing so.

The proper way of finding how long the earth takes to turn round is by observing the stars. Fix on any star we please, and note it in a certain position to-night; if you then note the moment when the star is in the same place to-morrow, the interval of time that has elapsed is the true duration of one complete rotation. That length is 23 hours 56 minutes 4 seconds, or about four minutes shorter than the ordinary day, measured from one noon to the next.

THE ANNUAL MOTION OF THE EARTH ROUND THE SUN.

I have as yet only been speaking of the *daily* movements by which the sun appears to go across the heavens between morning and evening; and we next consider the annual movements which give rise to the changes of the seasons. It is now Christmastide, when the days are short and dark, while six months ago the days were long and glorious in the warmth and brightness of summer. A similar recurrence of the seasons takes place every year, and thus we learn that some great changes take place every year in the relation between the earth and the sun. We must try and explain this. Why is it that we enjoy warmth at one part of the year, and suffer from frost and snow at another?

Note first a great difference between the sun in summer and the sun in winter. I will ask you to look out at noon any day when the clouds are absent, and you will then find the sun at the highest point it reaches during the day. All the morning the sun has been gradually climbing from the east; all the afternoon it will be gradually sinking down to the west. Let us make the same observation at different parts of the year. Suppose we take the shortest day in December. You will look out about twelve o'clock from some situation which affords a view towards the south, and there, as shown in the adjoining sketch (Fig. 22), is the mid-winter sun.

But now the spring approaches, and the days begin to lengthen. If you watch the sun you will see it pass higher and higher every noon until Midsummer Day is reached, and then the sun at noon is found in the upper position. As autumn draws near, the sun at noon creeps downwards again until, when the next shortest day has come round, we find that it passes just where it did at the previous mid-winter. With unceasing regularity year after year the sun goes through these changes. When he is high at noon we have days both long and warm; when he is low at noon we have days both short and cold.

Vulcan and his golden boat, of course, pretended to give an explanation of this. As the summer drew on, each day

Fig. 22.—The Changes of the Sun with the Seasons.

Vulcan shot out the sun with a stronger impulse, so that it should ascend higher and higher. His greatest effort was made on Midsummer Day, when, after rowing but a little way round from the north towards the east, he drove off the sun with a terrific effort. The sun soared aloft to the utmost height it could reach, and in the meantime Vulcan returned to the west to be ready to catch the sun as it descended. On the other hand, in mid-winter, he came round much further through the east to the south, and then shot up the sun with his feeblest effort, and had to paddle as hard as ever he could so as to complete his long return voyage during the brief day.

It is thus evident that there are two quite different kinds of motion of the sun. There is first the daily rising and setting, for which we have accounted by showing that it is merely an appearance produced by the fact that the earth is turning round. But now we have been considering quite a different motion by which the sun seems to creep up and down in the heavens, and this takes a whole year to go through its changes.

There is still another point which we must consider before we can understand all these puzzling movements of the sun. We shall ask the stars to help us by their familiar constellations. You know, perhaps, the Great Bear, or the Plough as it is often called, and Orion. There are also Aries the Ram, Taurus the Bull, and other fancifully named systems. These constellations have been known for countless ages, and for our present purposes we may think of them as permanent groups in the heavens, which do not alter either their own shapes or their positions relatively to each other. These groups of stars extend all around us. They are not only over our heads and on all sides down to the horizon, but if we could dig a deep hole through the earth coming out somewhere near New Zealand, and if we then looked through we should see that there was another vault of stars beneath us. We stand on our comparatively little earth in what seems the centre of this great universe of stars all around. It is true we do not often see the stars in broad daylight, but they are there nevertheless. The blaze of sunlight makes them invisible. A good telescope will always show the stars, and even without a telescope they can sometimes be seen in daylight in rather an odd way. If you can obtain a glimpse of the blue sky on a fine day from the bottom of a coal pit, stars are often visible. The top of the shaft is, however, generally obstructed by the machinery for hoisting up the coal, but the stars may be seen occasionally through the tall chimney attached to a manufactory when an opportune disuse of the chimney permits of the observation being made. The fact is that the long tube has the effect of completely screening from

Fig. 23.—How the Stars are to be seen in Broad Daylight.

the eye the direct light of the sun. The eye thus becomes more sensitive, and the feeble light from the stars can make their impression, and produce vision. From all these various lines of reasoning we see that there can be no doubt of the continuous presence of stars above and around us, and below us, on all sides, and at all times.

If you look out to-night, it being Christmas time, towards the south, you will see the Belt of Orion and the Dog Star in a most splendid portion of the heavens. These stars you will see every winter under similar circumstances. But you may look in vain for them in summer. Stars you can see in the summer evenings, but they will be totally different from those that adorned the skies in winter. Each season has its own constellations. This simple fact was known to the ancients, and we shall find its explanation full of meaning. I shall here name four constellations, being the best-known ones that answer our purpose. They lie in a circle round the heavens. They are Orion, Virgo, Scorpio, and Pisces. I am supposing that you are looking out at midnight towards the south. In December you will see Orion; in March, Virgo; in June, Scorpio; and in September, Pisces; and then next December you will be looking at Orion again. See what this proves. At midnight, of course, the sun is at the other side of the earth, so that if I am looking at Orion in mid-winter the sun must be behind my back. Look at our little picture (Fig. 24). The earth is in the middle, and the sun must be on the opposite side to Orion. That is, the sun must be somewhere about the position I have marked at A. In March we see Virgo in the south at midnight, when, of course, the sun is at the other side of the earth; so that the sun must be somewhere at B. In

June Scorpio is seen, so that the sun must be at the other side, at C. That is to say, in midsummer the sun is in that part of the sky where Orion is situated. If, therefore, on a bright June day we could see the stars we should find Orion in all his splendour in the south. But, of course,

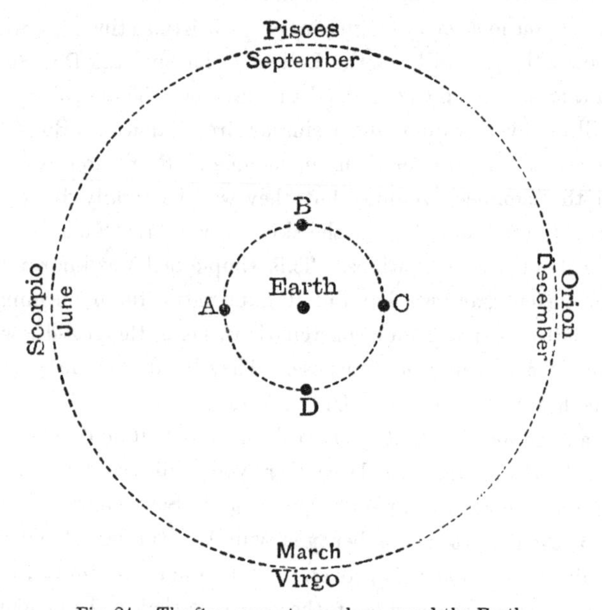

Fig. 24.—The Sun seems to revolve around the Earth.

the light of the sun makes Orion invisible. We can, however, see the stars by our telescopes, and on rare occasions an eclipse of the sun will occur, by which he is temporarily extinguished, and then we can see all the stars, even though it is daytime.

Thus it would seem as if the sun were first at A and

then at B, C, and D, and then began to go round again. I say it would *seem* as if the sun had these movements, and to the ancients there was no doubt about the matter. Even after it was plain that the earth turned round on its axis so as to give the changes of day and night, it was still thought necessary to suppose that the sun went round the earth once in the year, in order to explain how the changes in the stars during the different seasons were produced.

Here is another case in which we must be careful to distinguish between what appears to be true and what is actually true. Everything that we actually see would be just as well explained by supposing that the sun remained at rest, and that the earth revolved around it, as in Fig. 25. If, for instance, the earth were at A in mid-winter, then the sun is on the opposite side to Orion, and of course at midnight we shall be able to see Orion. So in spring the earth is at B, and we see Virgo, and similarly in summer we have Scorpio, and in autumn Pisces. Thus all that is actually seen could be fully accounted for by regarding the sun as fixed in the centre, and the earth as travelling round it from A to B, to C and to D respectively, and completing the journey in a twelvemonth. Which idea are we to adopt? Shall we say that the earth goes round the sun, or the sun goes round the earth?

I remember an old college story, which I cannot help giving you at this place. It may serve to lighten what I fear you must otherwise have thought rather a tedious part of our subject. There were three students brought up for examination in astronomy, and they showed a lamentable ignorance of the subject, but the examiner being a kind-hearted man wished, if possible, to pass them; and so he proposed to the three youths the very simplest question that he could think of. Accordingly, addressing the first student, he said: "Now tell me which, does the earth go round the sun, or the sun go round the earth?" "It is—the earth—goes round the sun." "What do you say," he

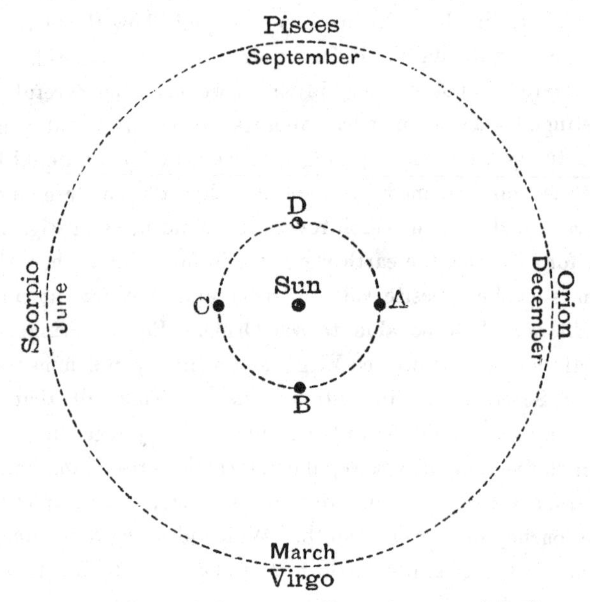

Fig. 25.—The Earth, however, really revolves around the Sun.

inquired, turning rather suddenly on the next, who gasped out: "Oh, sir—of course—it is the sun goes round the earth." "What do you say?" he shouted at the third unhappy victim. "Oh, sir, it is—sometimes one way, sir, and sometimes the other!"

But which is it? Well, we must remember that the

earth is comparatively a very little body and the sun a very big one, so it is not at all surprising to learn that the earth goes round the sun, which remains, practically speaking, at rest in the centre. Thus our great earth and all it contains are continually bound on an endless voyage round and round the sun. You will find it instructive to work out this little sum. How fast is the earth moving, or how far do we go in a second? We are about 93,000,000 miles from the sun, and the great circle that we go round has a diameter twice as great as this—that is, about 186,000,000 miles. The circumference of a circle is nearly three and one-seventh times its diameter, and accordingly the whole length of the voyage in the year is about 585,000,000 miles. This has to be accomplished in 365 days, so that the daily run must be about 1,600,000 miles. We divide this by 24, to find the distance journeyed each hour, which we find to be about 67,000 miles; and we must divide this again by 60 to find the length covered in a minute. It is truly startling to find that, night and day, this great earth has to travel for more than eighteen miles every second in order to get round its mighty path in the allotted time.

I began this lecture about forty minutes ago, and I think from what I have said you will be able to calculate a result that will, I dare say, astonish you. In these forty minutes we have moved about 45,000 miles. No doubt my lecture commenced in this hall, and in your presence; but can I truly say I began it *here?* Well, no; I began it not here, but at a place 45,000 miles away; but we have all been travelling together, and the journey has been so very smooth and free from all jolts that we never thought anything about the motion.

I am sure many of those to whom I am now speaking have read accounts of voyages in the Arctic regions. You have been told of the sufferings of the crews during the long winters, amid the ice and snow; and you have heard how, during that dismal period, there is total darkness, for the sun never rises for weeks and months together. On the other hand, these northern regions often present a more cheerful picture. During midsummer, the long darkness of winter is atoned for by perpetual sunshine. At midnight there is still the full brilliance of day, and the sun, though low no doubt, has not passed below the horizon. Even in the northerly parts of Europe, we can see the midnight sun. Lord Dufferin, in his delightful narrative of a cruise entitled "Letters from High Latitudes," gives an interesting illustration of the perplexities arising from endless daylight. It appears that everything went on happily until the fatal moment when the yacht crossed the Arctic circle. Then it was that dire tribulation arose among the poultry. A fine cock was the cause of the trouble. Knowing his duty, he always liked to be particular about performing the important task of crowing at sunrise. This he could do regularly, so long as the yacht remained in reasonable latitudes, where the sun behaved properly. But when they crossed the Arctic circle, the cock was confronted with a wholly new experience. The sun never set in the evening, and consequently never had the trouble of rising in the morning. What was the distracted bird to do? He did everything. He burst into occasional fits of terrific crowing at all sorts of hours, then he gave up crowing altogether, but finding that did not mend matters, he took to crowing incessantly. Exhaustion was succeeded by delirium, and rather than live any longer in a universe where the sun was capable of pranks so heartless, the indignant fowl flung himself from the vessel and perished in the Arctic Ocean.

THE CHANGES OF THE SEASONS.

In the adjoining figure, I show a little sketch (Fig. 26), by which I try to explain the changes of the seasons. It exhibits four positions of the earth, one on each side of the sun. The left, A, represents the earth when summer gladdens the northern hemisphere; while the right, C, shows winter in the same region. You will see the two central lines which represent the axis about which the earth rotates. Of course, the earth has no visible axis. That is an imaginary line which runs through the globe from the North Pole to the South Pole. It remains fixed in the earth, for we can prove in our observatories that the Pole does not shift its position to any appreciable extent in the earth itself. We know this by the fact that the latitudes of points on the earth remain practically constant. In fact, if we could reach the North Pole and drive a peg into the ground year after year to mark the exact spot, we should find that the position of the pole was sensibly the same. Does it not seem strange that we should be able to know so much about the Pole, though we have never been able to get there; have never, in fact, been able to get within less than 400 miles of it? I think you will be able to understand the point quite easily. The latitude of a place, as you know from your geography, is the number of degrees, and parts of a degree, between that place and the equator. In our observatories, we can determine this so accurately that the difference between the latitude of one

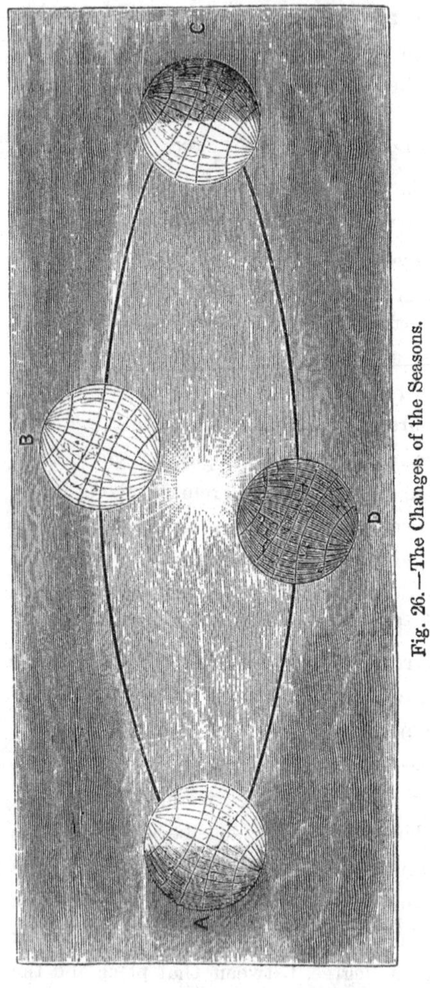

Fig. 26.—The Changes of the Seasons.

side of a room and of the other side of the same room is quite perceptible. As we find that the latitudes of our observatories remain sensibly unchanged from year to year, we are certain that the pole must remain in the same place. Indeed, if the pole were to alter its position by the distance of a stone's throw, the careful watchers in many observatories would speedily detect the occurrence.

And now I must direct your attention to something apparently quite different. When the battle of Waterloo was fought, the great victory was won with the aid of the old-fashioned musket, a smooth-bore gun, which was loaded at the muzzle with a good charge of powder, and then a round bullet was rammed down. "Brown Bess," as the musket was called, was a most efficient weapon at close quarters, and indeed at any distance *when the bullet hit;* but there was the difficulty. The round bullets rushing up the tube and out into the air anyhow, had a habit of roaming about, which was quite incompatible with accurate shooting.

One great improvement in small arms consisted in giving to the bullet a rapid rotation about an axis which is in the line of fire. This is what the *rifle* accomplishes. The grooves in the barrel of the rifle twist round, and though they only give half a complete turn in the length of the barrel, yet the speed of the bullet is so great that when it flies off it is actually spinning with the tremendous velocity of about one hundred and fifty revolutions a second. Even with the old-fashioned round bullet, the rifling of the barrel effected great improvement in the accuracy of the shooting. The introduction of the elongated bullets was another great improvement, while the adaptation of breech-loading enabled a bullet to be used rather larger than that which could have been forced down the barrel, and thus it was ensured that the grooves shall bite into the bullet as it hurries past, and impart the necessary spin.

A body rapidly rotating about an axis has a tendency to preserve the direction of that axis, and powerfully resists any attempt to change it. Our earth is spinning in this fashion. It is true that the rotation is, in one sense, a slow one, for it requires almost an entire day for each rotation. But when we remember the dimensions of our earth, we shall modify this notion. We have already stated that any place on the equator has to travel more than one thousand miles each hour, in order to accomplish the journey within the required time. So far, therefore, the earth moves like a rifle-bullet, and the direction of its axis remains constant.

In the course of the great voyage between summer and winter, the earth travels from one side of the sun to the opposite side, and in doing so it still continues to spin about an axis parallel to the original direction. See the consequences which follow. The sun illuminates half the earth, and in the left position in Fig. 26, representing summer, the North Pole is turned over towards the sun, and lies in the bright half of the earth. There is continual day at the North Pole, and night is unknown there at this time of year, because the turning of the earth about its axis will not bring the Pole nor the regions near the Pole into the dark hemisphere. Thus it is that the Arctic regions enjoy perpetual day at this season. Look now at the position of England when the northern hemisphere of the earth is tilted towards the sun, and is consequently enjoying the full splendour of midsummer. As the earth turns round, England will gradually cross over the boundary between light and shade, and will enter the dark part. Then there will be night in England, but you will see from the figure that the day is much longer than the night, and hence it is that we enjoy the fine long days in summer.

We next look at a different scene six months later. The earth has reached the other side of the sun, but the axis has remained parallel to itself, consequently the North Pole is now turned entirely away from the sun. The earth continues to turn round as before, but its movements do not bring the North Pole nor the surrounding Arctic regions out of the dark hemisphere, and consequently the night must be unbroken in these dismal circumstances. The long continuous day which forms the Polar midsummer is dearly purchased by the gloom and cold of a winter in which there is no sun for many weeks in succession. Observe also the changed circumstances of England. In the course of each twenty-four hours it lies much longer in the dark half of the earth than in the bright, and consequently there is only a short day succeeded by a long night.

SUNSHINE AT THE NORTH POLE.

It is the privilege of astronomers to be able to predict events that will happen in thousands of years to come, and to describe things accurately though they never saw them, and though nobody else has ever seen them either. No one has ever yet got to the North Pole, but whenever they do, we are able to tell them much of what they will see there. I must leave it to Jules Verne to describe how the journey is to be made, and how the party are to be supported at the North Pole. I shall give a picture of the changes of the seasons, and of the appearances in the stars, as seen from thence.

We shall, therefore, prepare to make observations from that very particular spot on this earth—the North Pole. I suppose that eternal ice and snow abides there. I don't think it would be a pleasant residence, however we shall arrange to arrive on Midsummer Day, prepared to make a year's sojourn. The first question to be settled is the erection of the hut. In a cold country it is important to give the right aspect, and we are in the habit of saying that a southerly aspect is the best and warmest, while the north, or the east, are suggestive only of chills and discomfort. But what is a southerly aspect at the North Pole, or, rather, what is not a southerly aspect? Whatever way we look from the North Pole we are facing due south. There is no such thing as east or west; every way is the southward way. This is truly an odd part of the earth. The only other locality at all resembling it would be the South Pole, from which all directions would appear to be north.

The sun would be moving all through the day in a fashion utterly unlike any similar performance elsewhere. There would, of course, be no such thing as rising and setting. The sun would, indeed, at first seem neither to go any nearer to the horizon nor to rise any higher above it, but would simply go round and round the sky. Then it would gradually get lower and lower, moving round day after day in a sort of spiral, until at last it would get down so low that it would just graze the horizon, right round which it would circulate till half the sun was below, and then until the whole disk had disappeared. Even though the sun had now vanished, a twilight glow would for some time be continuous. It would seem to move round and round from a source below the horizon, then gradually the light would become fainter and fainter until at last the winter of utter and continuous blackness had set in. The first indications of the return of spring would be detected by a feeble glow near the horizon, which would seem to move round and round day after day. Then this glow would pass into a continuous dawn, gradually increasing until the sun's edge crept into visibility, and the great globe would at last begin to climb the heavens by its continual spiral until midsummer was reached, when the change would go on again as before.

Our first excursion to the country of Star-land has now been taken, and we have naturally commenced by studying that sun to which we owe so much. But we shall have to learn that though our sun is of such vital importance to us, yet, in magnificence and size, he has many rivals among the host of stars.

LECTURE II.

THE MOON.

The Phases of our Attendant the Moon—The Size of the Moon—How Eclipses are Produced—Effect of the Moon's Distance on its Appearance—A Talk about Telescopes—How the Telescope Aids us in Viewing the Moon—Telescopic Views of the Lunar Scenery—On the Origin of the Lunar Craters—The Movements of the Moon—On the Possibility of Life in the Moon.

THE PHASES OF OUR ATTENDANT THE MOON.

THE first day of the week is related to the greatest body in the heavens—the sun—and accordingly we call that day Sun-day. The second day of the week is similarly called after the next most important celestial body—the moon—and though we do not actually say Moon-day, we do say Monday, which is very nearly the same. In French, too, we have *lune* for moon, and *Lundi* signifies our Monday. The other days of the week also have names derived from the heavens, but of these we shall speak hereafter. We are now going to talk about the moon.

We can divide the objects in this room into two classes. There are the bright faces in front of me, and there are the bright electric lights above. The electric lights give light, and the faces receive it. I can see both lights and faces; but I see the lights by the light which they themselves give. I see the faces by the illumination which they have received from the electric lights. This is a very simple distinction, but it is a very important one in Star-land. Among all these bodies which glitter in the heavens there are some which shine by their own light like the lamps. There are others only brilliant by reflected light like the faces. It seems impossible for us to confuse the brightness of a pleasant face with the beam from a pretty lamp, but it is often not very easy to distinguish in the heavens between a body which shines by its own light and a body which merely shines by some other light reflected from it. I think many people would make great mistakes if asked to point out which objects on the sky were really self-luminous and which objects were merely lighted up by other bodies. Astronomers themselves have been sometimes deceived in this way.

The easiest example we can give of bodies so contrasted is found in the case of the sun and the moon. Of course, as we have already seen, the sun is the splendid source of light which it scatters all around. Some of that light falls on our earth to give us the glories of the day; some of the sunbeams fall on the moon, and though the moon has itself no more light than earth or stones, yet when exposed to a torrent of sunbeams, she enjoys a day as we do. One side of her is brilliantly lighted; and this it is which renders our satellite visible.

Hence we explain the marked contrast between the sun and the moon. The whole of the sun is always bright; while half of the moon is always in darkness. When the bright side of the moon is turned directly towards us, then, no doubt, we see a complete circle, and we say the moon is full. On other occasions a portion only of the bright surface is turned towards us, and thus are produced the beautiful crescents and semi-circles and other phases of the moon.

We have here a simple apparatus (Fig. 27) which will explain their various appearances. The large india-rubber ball there shown represents the moon, which I shall illuminate by a beam from the electric light. The side of the ball turned towards the light is glowing brilliantly, and from the right side of the room you see nearly the whole of

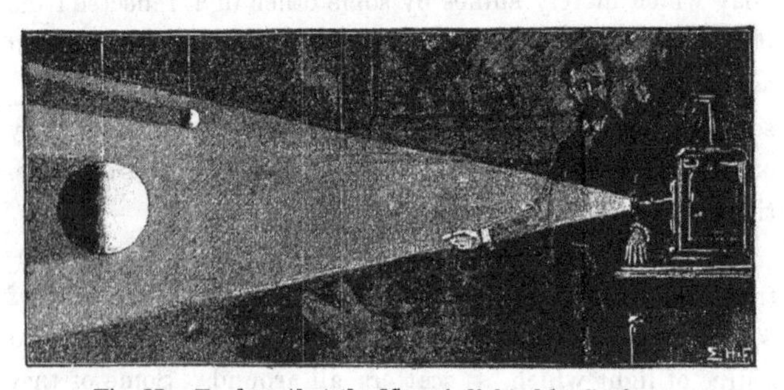

Fig. 27.—To show that the Moon is lighted by Sunbeams.

the bright side. To you the moon is nearly full. From the centre of the room you see the moon like a semi-circle, and from the left it appears a thin crescent of light. I alter the position of the ball with respect to the lamp, and now you see the phases are quite changed. To those on my left our mimic moon is now full; to those on my right the moon is almost new, or is visible with only a slender crescent. From the centre of the room the quarter is visible as before. We can also show the same series of changes by the little contrivance of Figs. 28 and 29.

Thus every phase of the moon (Fig. 30), from the thinnest beautiful crescent of light that you can just see low in the west after sunset up to the splendour of the full moon, can be completely accounted for by the different aspects of a globe, of which one half is brilliantly illuminated.

We can now explain a beautiful phenomenon that you will see when the moon is still quite young. We fancifully describe the old moon as lying in the new moon's arms, when we observe the faintly illuminated portion of the rest of that circle, of which a part is the brilliant crescent. This can only be explained by showing how some

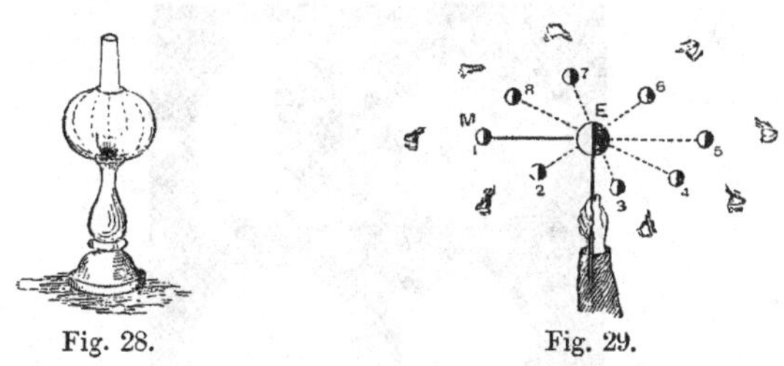

Fig. 28. Fig. 29.

The Phases of the Moon.

light has fallen on the shadowed side; for nothing which is not itself a source of light can ever become visible, unless illuminated by light from some other body.

Let me suppose, just for this moment, that there is a man on the moon, who is looking at the earth. To him the earth will appear in the same way as the moon appears to us, only very much larger. At the time of new moon the bright side of the earth will be turned directly towards him, so that the man in the moon will see an earth nearly full, and consequently pouring forth a large flood of light. Think of the brightest of all the bright moonlight nights you have ever seen on earth, and then think of a light which would be produced if you had thirteen moons, all as big and as bright as our full moon, shining together. How splendid the night would then be! you would be able to read a book quite easily! Well, that is the sort of illumination which the lunar man will enjoy under these circumstances; all the features of his country will be brightly lighted up by the full earth. Of course, this earth-lighted side of the moon cannot be compared in brilliancy with the sun-lighted side, but the brightness will

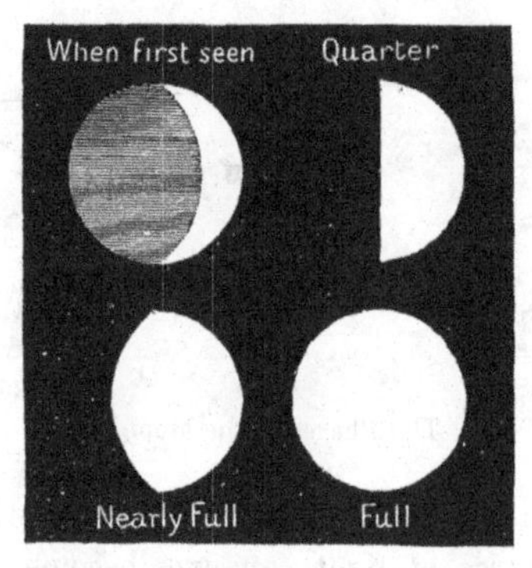

Fig. 30.—The Changes in the Moon.

still be perceptible, so that when from the earth we look at the moon, we see this glow distributed all over the dark portion; that is, we observe the feebly lighted globe clasped in the brilliant arms of the crescent. At a later phase the dark part of the moon entirely ceases to be visible, and this for a double reason: firstly, the bright side of the earth is then not so fully turned to the moon, and therefore the illumination it receives from earthshine is not so great; and secondly, the increasing size of the sun-lighted part of the moon has such an augmented glow that the fainter light is overpowered by contrast. You must remember that more light does not always increase the number of things that can be seen. It has sometimes the opposite effect. Have we not already mentioned how the brightness of day makes the stars invisible? The moon herself, seen in full daylight, seems no brighter than a shred of cloud.

THE SIZE OF THE MOON.

It is not easy to answer the question which I am sometimes asked, "Is the moon very big?" I would meet that question by another, "Is a cat a big animal?" The fact is, there is no such thing as absolute bigness or smallness. The cat is no doubt a small animal when compared with the tiger, but I think a mouse would probably tell you that the cat was quite a big animal, rather too big indeed, in the mouse's opinion. And the tiger himself is small compared with an elephant, while the mouse is large as compared with a fly.

When we talk of the bigness or the smallness of a body, we must always consider what we are going to compare it with. It is natural in speaking of the moon to think of it beside our own globe, and then we can say that the moon is a small body.

The relative sizes of the earth and the moon may be illustrated by objects of very much smaller dimensions. Both a tennis-ball and a foot-ball are no doubt familiar objects to everybody by this time. If the earth be represented by the foot-ball, then the moon would be about as large as the tennis-ball. But this proportion is not quite accurate, so I will suggest to you an interesting way of making a better pair of models of the earth and the moon.

In fact, experiments somewhat similar to those I describe have been actually going on in every kitchen in the land during this festive season. For have not globes and balls of all sorts and sizes been made of plum-pudding, and it will only require a little care on the part of the cook to make a pair of luscious puddings that shall fairly set forth the sizes of the earth and the moon. There is first to be a nice little round plum-pudding, three inches in diameter.

Fig. 31.—Relative Sizes of the Earth and the Moon.

It is just a little bigger than a cricket ball. It should, however, only make its appearance at a bachelor's table. Were it set down before a hearty circle on Christmas Day dire disappointment would result. One boy of sound constitution could eat it all. Perhaps it would weigh about three-quarters of a pound. This little globe is to represent the moon.

Another plum-pudding is to be constructed, which shall represent the earth (Fig. 31). We must, however, beg the cook to observe the proportions. The width of the earth, or the diameter, to use the proper word, is about four times the diameter of the moon. Hence, as the small plum-pudding was three inches across, the large one must have a diameter of twelve inches. This will be a family pudding of truly satisfactory dimensions; perhaps the cook will be a little surprised to find the alarming quantity of materials that will be required to complete a sphere of plum-pudding a foot in diameter.

These models having been duly made, and boiled, and placed on the table, we are now to propose the following problem :—

"If one school-boy could eat the small plum-pudding, how many boys would be required to dispose of the large one?"

The hasty person, who does not reflect, will at once dash out the answer, "Four!" He will say, "It is quite plain that, since one of the puddings has four times the diameter of the other, it must be four times as big; and therefore, as one boy is able to eat the small pudding, four boys will be adequate for the large one." But the hasty person will, as usual, be quite wrong. His argument would be sound if it were merely two pieces of sugar-stick that he was comparing; no doubt there is only four times as much material in a piece twelve inches long, as there is in a piece three inches long. But the plum-puddings have breadth and depth, which are in the same proportions as the length, and the consequence is that the large plum-pudding is far more than four times as big as the small one. No four boys, however admirable their capacities, would be equal to the task of consuming it. Nor even if four more boys were called in to help would the dish be cleared. Twenty boys, forty boys, fifty boys would not be enough.

It would take sixty-four boys to demolish the magnificent plum-pudding one foot in diameter.

If the cook will try the experiment, she will find that by taking the materials sufficient for sixty-four small plum-puddings all of the same size, and mixing them together, she will, no doubt, make a large plum-pudding, but its diameter will only be four times that of the small puddings.

As a matter of fact, the moon is 2,160 miles in diameter, and the earth is 7,918 miles. These numbers are so nearly 2,000 and 8,000 respectively, that for simplicity I have spoken of the earth as having a diameter four times as great as the moon. If we want to be very accurate, we ought to determine the ratio of the two quantities from the figures just given. Our illustration of the plum-puddings must, therefore, be a little modified. The earth is not quite so much as sixty-four times as big as the moon; but we need not now trouble about this point.

Another interesting question may be proposed, namely: How much land is there on the moon? We might state the answer in acres or in square miles; but it will, perhaps, be more instructive to make a comparison between the moon and the earth.

Here also I shall use an illustration; and we shall again consider two globes which are respectively three inches and twelve inches in diameter. The globes I use this time are hollow balls of india-rubber. These will represent the earth and the moon with sufficient accuracy, and the relative surfaces of these two globes is what I want to find. There are different ways in which the comparison might be made. I might, for instance, paint the two globes and see the quantity of paint that each requires. If I did this, I should find that the great globe took just sixteen times as much paint as the small one. We can adopt a simpler plan. The india-rubber in one of these balls has the same thickness as in the other, so that the quantity which is required for each ball may be taken to represent its surface. By simply weighing the two balls, I perceive that the large one is sixteen times as heavy as the small one. You notice here the difference between the comparative weights of two hollow balls and two solid ones of the same size and the same material. Had these globes been of solid india-rubber, the large one would have weighed sixty-four times as much as the small one, just as in the case of the plum-puddings; but being hollow, the ratio of their weights is only the square of the ratio of their diameters—that is to say, four times four, or sixteen.

We are thus taught that if the moon were exactly one-fourth of the diameter of the earth, its surface would be one-sixteenth part of that of the earth. It would, no doubt, have made our subject a little easier and simpler if the moon had been created somewhat smaller than it is. As, however, the universe has not been solely constructed for the purpose of these talks about Star-land, we must take things as we find them. This proportion is not four; it is more nearly $3\frac{2}{3}$, and the relative surfaces of the two bodies is the square of $\frac{11}{3}$, or about $13\frac{1}{2}$. In other words, the entire extent of our globe is about thirteen and a half times the extent of the moon.

The face of the full moon being half the entire extent of the surface is, therefore, about one twenty-seventh part of the earth's surface—continents, oceans, seas, and islands all taken together. The British Empire and the Russian Empire are each of them as large as the face of the full moon.

HOW ECLIPSES ARE PRODUCED.

The moon is the attendant, or the satellite of the earth, ministering to the wants of the earth by mitigating the

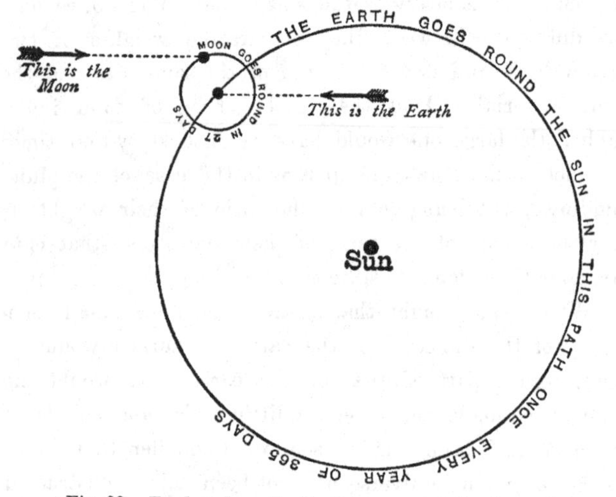

Fig. 32.—To show how the Earth goes round the Sun and the Moon goes round the Earth.

darkness of our nights. The earth goes around the sun in its annual journey of 365 days. The moon revolves around the earth once every twenty-seven days. The motion of the moon is thus a very complicated one, for it is actually moving round a body which is itself in constant motion (Fig. 32).

You will see by your almanacs every year that certain

eclipses are to take place; and after what we have said about the sun and the moon, it will be easy to understand how eclipses arise. There are two different kinds. You will sometimes see an eclipse of the moon, and sometimes those eclipses of the sun of which we have spoken in the last Lecture. You may be surprised to find with what accuracy the eclipses can be predicted. We can tell not only those that will occur this year and next year, but we could also foretell the eclipses that will appear in a hundred or a thousand years to come; or we can, with equal ease, calculate backwards, so as to find the circumstances of eclipses that happened thousands of years ago. This shows how well we have learned the way the moon moves.

An eclipse of the sun is the simpler occurrence, so we shall describe it first. It happens when the moon comes between the earth and the sun. Look at our little astronomers, shown in Fig. 33. A boy and a girl are both gazing at the sun, when the moon comes between. To the boy the moon appears to take a great bite out of the sun, so that it looks like the left-hand picture in Fig. 34. (I have drawn a line from the end of the telescope in Fig. 33, which shows how much of the sun is cut off.) This would be called a partial eclipse of the sun. The almanac will sometimes describe the eclipses as visible in London, or visible at Greenwich; but that need not be taken so literally as was supposed by a Kensington gentleman, who, on noticing that the almanac said an eclipse was to be visible in London, called a cab and drove into the City to look for it. His almanac had not mentioned that it would be visible from his own house. You may usually take for granted that when an eclipse is said to be visible from London or

Greenwich, it will be more or less visible all over England Most of these eclipses are only partial, and though they are

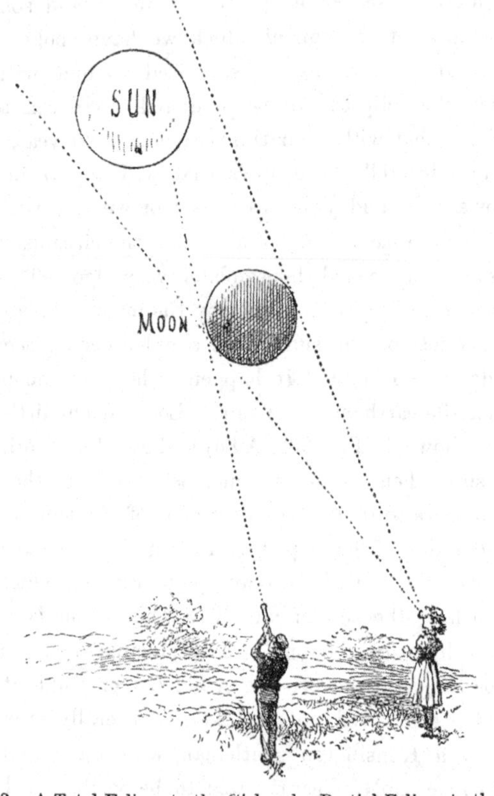

Fig. 33.—A Total Eclipse to the Girl and a Partial Eclipse to the Boy.

interesting to watch they do not teach us much. By far the most wonderful kind of eclipse is that in which the whole of the bright part of the sun is blotted out. Then, indeed, we do see wonders. But such eclipses are very rare, and even when they do occur they only last a very few minutes. The sights that are displayed are so interesting that astronomers often travel thousands of miles to reach a suitable locality for making observations.

The girl in Fig. 33 is placed in the best possible position for seeing the eclipse. There you find her right in the

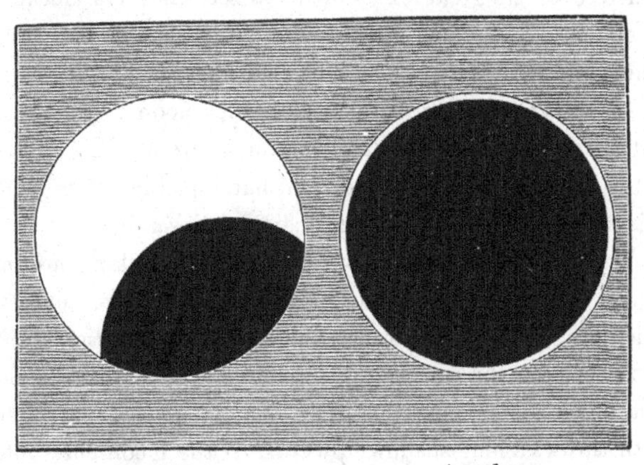

Partial. Annular.

Fig. 34.—Different kinds of Solar Eclipse.

line of the sun and moon; and I think you will agree that she cannot see any part of the sun, for the moon is altogether in the way. I have drawn two dotted lines, one at each side. All that she can see must lie outside these dotted lines, and she will be in the dark as long as the moon stays in the way. When the eclipse is complete comparative darkness steals over the land. The birds are deceived, and fly home to the trees to roost. The owls and the bats, thinking their time has arrived, venture forth on their nocturnal business. Even flowers close their petals, only to open a few minutes later when the sun again bursts forth. Other flowers that give forth their fragrance at night are also sweetly perceptible so long as the sun remains obscured. An unruly cow, accustomed to break into a meadow at night, was found there after an eclipse was over; while I learn from the same authority that a man rushed over in great excitement to see what his chickens were doing, but came back much disappointed on finding them pecking away as if nothing had happened.

It will sometimes happen that the moon is so placed that the edge of the sun can be seen all round it. A case of the kind is shown in the right-hand picture of Fig. 34. It is called an annular, or ring-shaped eclipse.

The eclipses of which we have been speaking are, of course, only to be seen during the day. The lunar eclipses, which are visible at night, are due to the interposition of the earth between the sun and the moon. The sun is at night time under our feet at the other side of the earth, and the earth throws a long shadow upwards. If the moon enter into this shadow, it is plain that the sunlight is partly or wholly cut off, and since the moon shines by no light of her own, but only by borrowed light from the sun, it follows that when the moon is buried in the shadow all the direct light is intercepted, and she must lose her brilliancy. Thus we obtain what is called a lunar eclipse. It is total if the moon be entirely in the shadow. The eclipse is partial if the moon be only partly in the shadow. The lunar eclipse is visible to everybody on the dark hemisphere of the earth if the clouds will keep out of the way, so that usually a great many more people can see a lunar eclipse than a solar eclipse, which is only visible from a limited part of the earth. It thus happens that the lunar eclipse is the more familiar spectacle of the two.

If the moon were entirely buried in the shadow, one might naturally think that it would become totally invisible. This is not always the case. It is a curious fact that in the depth of a total eclipse the moon is often still visible, for she glows with a copper-coloured light, which is bright enough to render some of the chief marks on her surface discernible.

EFFECT OF THE MOON'S DISTANCE ON ITS APPEARANCE.

We are going to take a good look at the moon and examine the different objects on it as well as we can. There is a peculiar interest attached to our satellite, because it is much the nearest of all the heavenly bodies to our globe, and therefore the one that we can see the best. Every other object—sun, star, or planet—is hundreds, or perhaps thousands, of times as far off as the moon. It is right that we should desire to learn all we can about the bodies in space. We know that our earth is a great ball, and we see that there are many other such bodies. Some of them are much larger, and some of them are smaller; some of them are dark globes like the earth, and among them the moon is one. Is it not reasonable that we should make special efforts to find out all we can about our neighbour?

Though the moon is so close to us when its distance is compared with that of other objects in space, yet when we express its distance in ordinary units it is a very long way off—about 240,000 miles—a length nearly as great as that of all the railways in the world put together. An express train which runs forty miles an hour would travel 240 miles in six hours, and the whole distance to the moon would be accomplished in 6,000 hours, so that travelling by night and day incessantly you would accomplish the journey in 250 days. To take another illustration, if you wrapped a thread ten times round the equator of the earth, it would be long enough to stretch from the earth to the moon. Or suppose a cannon could be made sufficiently strong to be fired with a report loud enough to be audible 240,000 miles away. The sound would only be heard at that distance a fortnight after the discharge had taken place.

The moon is too far for us to examine the particular features on its surface by the unaided eye. Suppose that there was a mighty city like London on the moon, with great buildings and teeming millions of people, and you went out on a fine night to take a look at our neighbour. What do you think you would be able to see of the great lunar metropolis? Would you be able to see its streets full of omnibuses, or even its great buildings? Would you see St. Paul's and Westminster—the great parks and the river? Of all these things your unaided eye would show you almost nothing. I can give you a little illustration. Suppose that you made a tiny model of London; imagine this little structure all complete, so that the streets, the buildings, the bridges, the railways, the parks, and the Thames were placed in their true proportions; suppose that the miniature city was so small that it could stand on a penny postage stamp, surely everything would look very insignificant, even if you had the model in your hand and looked at it with the aid of a magnifying glass. But suppose it were put on the other side of the table or on the other side of the room, or the other side of the street. Even St. Paul's Cathedral itself would have ceased to be distinguishable; but yet the distance is not nearly great enough. You would have to put the little model a quarter of a mile away before it would be in the right position to illustrate the appearance of a lunar London to the unaided eye.

A TALK ABOUT TELESCOPES.

The astronomer will not be contented with a mere naked eye inspection of a world so interesting as the moon. He will get a telescope to help his vision. The word "telescope" means a contrivance for looking at objects which are a long way off. We have explained that the further an object is the smaller it appears to be. The telescope enables us in a great degree to neutralise this inconvenience. It has the effect of making a distant object look larger.

There are great differences in the forms of telescopes; and some instruments are large and some small, according to the purposes for which they are required. Perhaps the most useful practical application of the telescope is by the officer on duty on board a ship. He is generally provided with a pair of these instruments bound together to form the "binocular."

You are all acquainted with this useful contrivance, or at all events with the opera-glass, that is used for purposes with which landsmen are more familiar. The ship's telescope, or the binocular, or the opera-glass, are feeble in power, when compared with the great instruments of the Observatory. The officer on the ship will generally be satisfied with a telescope which shall show the objects with which he is concerned at about one-third of their actual distance. Thus, suppose his attention is directed to a great steamer three miles away, he wishes to see her better, and accordingly he takes a view through his binocular. Immediately the vessel is transformed so that it seems to be only one mile away. The apparent dimensions of the object are increased three-fold. The hull is three times as long, the masts and the funnel are three times as high, the sailors are three times as tall; various objects on the ship too small to be seen at three miles would be visible from one mile, and to that apparent distance the ship has now been brought.

If the sailor desires to obtain a means of reducing the apparent distance of objects, how much more keenly does the astronomer feel the same want! At best, the sailor only has to scan a range of a few miles with his glass, but what are a few miles to the astronomer? It is true that he can count the distance of the moon by thousands of miles, a good many thousands, no doubt, but for all other objects he must use millions, while for most bodies in space, millions of millions of miles are the kind of figures we are constrained to employ. Need it be said that the astronomer must resort to every device by which he can reduce this apparent distance? He does not despise the modest binocular. It is often a useful instrument in the observatory. It gives most beautiful pictures of the celestial scenery, and you would be amazed to find how many thousands of stars you can see with it, that your unaided eye would not show you at all. The binocular will also greatly improve the appearance of the moon, but still its powers fall far short of what we require for the study of lunar landscapes. Even though we can reduce the moon's apparent distance to one-third its actual amount, yet still that third is a very considerable quantity. One-third of 240,000 is 80,000, so that we can see the moon no better with a binocular than we should see it were it 80,000 miles away, and were we viewing it with the unaided eye.

I am not going to enter here upon any detailed account of the telescope, because I shall say a little more on the subject in a later lecture; at present I only describe that form of instrument which is most convenient for studying the moon. I take as an illustration the South Equatorial at Dunsink Observatory, which belongs to Trinity College, Dublin.

This telescope has a building to itself, which stands on the lawn in front of the house. The site is open and elevated, so as to command a clear view all round. You will see in Fig. 35 a picture of the structure. It is circular in form and is entered by the little porch. The most peculiar feature of an edifice intended to contain this kind of telescope is its roof, or *Dome* as we call it. It is of an hemispherical shape with a projecting rim at the bottom. But no one would go to the trouble and expense of making a round dome like that over the Observatory, if it were not necessary for a particular purpose. The dome is very unlike ordinary roofs, not only in appearance, but also because it can turn round. In the next figure you will see a section through the building, and the wheels are exposed by which the dome is carried. These wheels run easily on rails, so that when the attendant pulls the rope which you see in his hands, he turns round a large pulley, and that turns a little cog-wheel which works into a rack, and thus makes the dome revolve. The roof is built of timber, covered with copper; it weighs more than six tons, but the machinery is so nicely adjusted, that a child four years old

Fig. 35.—The Dome at Dunsink Observatory.

can easily set the whole in motion. The object of all this machinery is seen when we learn that there is only one opening in the dome. It is covered by the shutter shown over the doorway in Fig. 35. When opened to the top, it gives a long and wide aperture, through which the astronomer

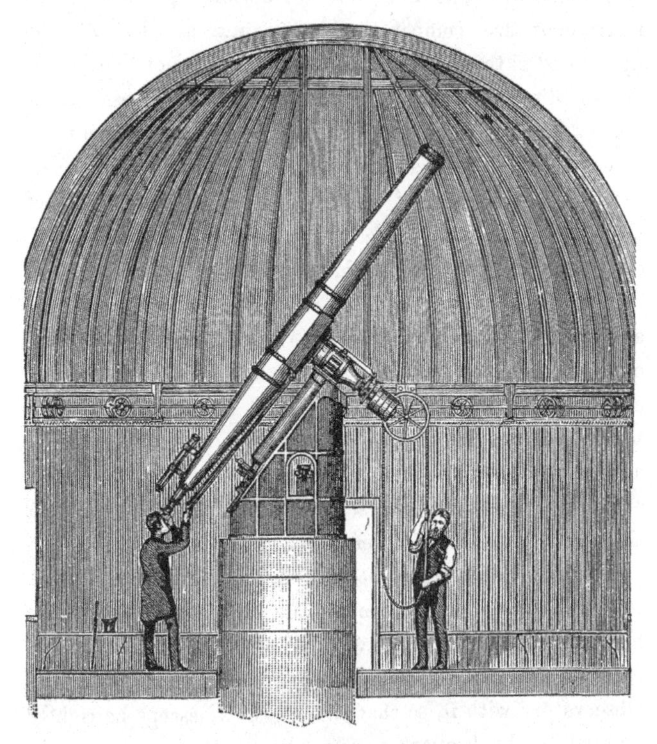

Fig. 36.—The Equatorial at Dunsink.

can look out at the heavens. Of course the dome has to be turned until the opening has been brought to face the required aspect. The big telescope can thus be directed to any object above the horizon. You see a gentleman using the telescope (Fig. 36), and this shows that the great instrument is nearly three times as long as the astronomer himself! No doubt the telescope seems to be composed of a good many different parts, but the essential portions of the instrument are comparatively few and simple. At the upper end is the object glass, which consists of two lenses, one of flint glass and the other of crown glass. These glasses must be of exceptional purity, and the shape to be given to the lenses is a matter of the utmost importance. It is in the making of this pair of lenses that the skill of the optician has to be specially put forth. So valuable indeed is an objective which fulfils all the requirements, that it is by far the most costly part of the instrument. There are no further glasses in the interior of the tube until you come to the end where the observer is looking in. This is closed by an eye-piece consisting of a lens, or a pair of lenses. There are usually many eye-pieces to a telescope, and they contain lenses of different powers, to be used according to the various states of the atmosphere, or to the particular kinds of observation in progress.

If you point a big telescope to the sky, and see therein the sun or the moon or any of the stars, you will speedily find that the objects pass away out of view. Remember our earth is constantly turning round, and bears, of course, the Observatory with it, so that though the telescope be rightly pointed to the heavens at one moment, by the next it will have turned aside. To you who are using the telescope, the appearance produced is as if the heavenly bodies were themselves moving. We can counteract this inconvenience. The telescope is supported on a pedestal, which is built on masonry, that goes down through the floor to its foundation on the solid rock beneath. In the iron casing at the the top of the pedestal you will see a little window, and inside there is clockwork driven by a heavy weight. This clockwork turns the whole telescope round in the opposite direction to that in which the earth is moving it. The consequence is that the telescope remains constantly pointed to the same part of the heavens. You will thus find that the stars or planets remain apparently stationary in the field of view when once the clock has been set in motion.

This instrument is no doubt a large one, but of late years many much greater have been built. There are some nearly twice as long and more than twice as wide.

HOW THE TELESCOPE AIDS US IN VIEWING THE MOON.

Those who are in charge of an observatory are often visited by persons who, coming to see the wonders of the heavens, and finding instruments of such great proportions, not unnaturally expect the views they are to obtain of the celestial bodies shall be of corresponding magnificence. So they are, no doubt, but then it frequently happens that the pictures which even the greatest telescope can display, will fall far short of the ideal pictures which the visitors have conjured up in their own imaginations, so that they are often sadly disappointed. Especially is this true with regard to the moon. I have seen people who, when they had a view of the moon through a great telescope, were surprised not to find vast ranges of mountains which looked to them as big as the Alps, or mighty deserts, over which the eye could roam for thousands of miles. They have sometimes expected to behold stupendous volcanoes, that not only were, but that looked to be as big as Vesuvius. Others seem to have thought they ought to see the moon with such clearness that the fields were to be quite visible, and some would not have been much astonished if they had observed houses and farmyards, and, perhaps, even cocks and hens.

There are different ways of estimating the apparent dimensions of an object, but the size the moon appears to me to have in a great telescope may be illustrated by taking an orange in your hand and looking at the innumerable little marks and spots on its surface. The amount of detail that the eye will show on the orange is about equal to the amount of detail that a good telescope will show on the moon. A desert on the moon, which really is a hundred miles across, will then correspond to a mark about an eighth of an inch in diameter on the orange. Some of you may ask what is gained by the use of a telescope, for the moon looks to us as large as a plate with the unaided eye, and now we hear it only looks as big as an orange in the telescope. But where is the plate with which you compare your moon supposed to be held? It is surely not in your hand. It is imagined to be up in the sky, a very long way off. Though an orange is much smaller than a plate, yet you will be able to see many more details in the orange by taking it in your hand than you could see on a plate which was at the other side of the street.

I sometimes find that people will not believe how much the telescope that they are using is magnifying the moon until they use both eyes together, of which one is applied to the telescope, while the other is directed to the moon (Fig. 37). It will then be seen, even with a very small instrument, that the telescopic moon is as big as the larger of the two crescents in the adjoining figure, while the naked eye moon is like the smaller.

The greatest of all telescopes is capable of reducing the apparent distance of an object to about one-thousandth part of its actual amount. If, therefore, a body were a

Fig. 37.—Demonstration of the Advantage of Using a Telescope.

thousand miles away, it would, when viewed by one of these mighty instruments, be seen as large as our unaided vision would show it were the body only a single mile distant. No doubt this is a large accession to our power, but it often falls far short of what the astronomer would desire. The distances of the stars are all so great that even when divided by one thousand, they are still enormous. If you have a number expressed by 100,000,000,000,000, then dividing it by a thousand merely means taking off three of the cyphers, and there are still a large number left. We are, however, at present concerned with the moon, and, as its distance is about 240,000 miles, the effect of the best telescope is to reduce this distance apparently to 240 miles. Here, then, we find a limit to what the best of all telescopes can do. It can never show us the moon better, hardly indeed so well as we could see it with our unaided eye were it only 240 miles over our heads. We cannot expect the most powerful instruments to reveal any object on the moon unless that object were big enough to be seen by the unaided eye when 240 miles away. What could we expect to see at a distance of 240 miles?

Here is a little experiment which I made to study this point. I marked a round black dot on a sheet of white paper. The dot was a quarter of an inch in diameter, and then I fastened this on a door in the garden, and walked backwards until it ceased to be visible. I found this distance to be about thirty-six yards. I tried a little boy of eight years old, and it appeared that the dot became invisible to him about the same time as it did to me. What has this to do with the moon? you will say. Well, we shall soon see. In thirty-six yards there are 5,184 quarters of an inch, and as it is unnecessary to be very particular about the figures, we may say, in round numbers, that the distance when we ceased to be able to distinguish the dot was about five thousand times as great as the width of the dot itself. You need not, therefore, expect to see anything on the moon which is not at least as wide as the five-thousandth part of the distance from which we are viewing it. The great telescope practically places the moon at a distance of 240 miles, and the five-thousandth part of that is about eighty yards; consequently a round object on the moon about eighty yards in diameter would be just glimpsed as the merest dot in the most powerful telescope. To attract attention, a lunar object should be much larger than this. If St. Paul's Cathedral or a rock of that size stood on a lunar plain, it would be visible in our great telescopes. It is true that we could not see any details. We should not be able to distinguish between a Cathedral and a Town-hall. There would just be something visible, so that the artist who was making a sketch of that part would make a mark with his pencil to show that something was there. This will show us that we need not expect to see objects on the moon, unless they are of great size, even with the mightiest of telescopes.

TELESCOPIC VIEWS OF THE LUNAR SCENERY.

But we must hasten to make our telescopic scrutiny of the moon. I have already warned you not to expect too much, even with the biggest of telescopes; and just as a caution, I may, perhaps, tell you a story I once heard of an astronomer who had a great telescope. It was a very famous instrument, and people often came to the Observatory at night to enjoy a look at the heavens. Sometimes these visitors were grave philosophers, but frequently they were not very accomplished men of science. One evening such a visitor came to the Observatory, and sent in his name and an introduction to the astronomer, with a request that he might enter the temple of mystery. The astronomer courteously welcomed the stranger, and asked him what he specially desired to see.

"Oh!" said the visitor, "I have specially come to see the moon—that is the object I am particularly interested about."

"But," said the astronomer, "my dear sir, I would show you the moon with pleasure, if you were here at the proper time; but what brings you here now? Look up; the evening is fine. There are the stars shining brightly, but where is the moon? You see it is not up at present. In fact, it won't rise till about half-past two to-morrow morning, and it is only nine o'clock now. Come back again in five or six hours, and you shall observe the moon with the great telescope."

But the visitor evidently thought the astronomer was merely trying to get rid of him by a pretext. And he was equal to the occasion—he was not going to be put off in that way.

"Of course, the moon is not up," he replied; "any one can see that, and that is the reason why I have come, for *if the moon had been up, I could have seen it without your telescope at all!*"

Although no land surveyor has ever yet been able to reach our satellite, yet it is hardly an exaggeration to say that in some respects we know the geography of the moon a good deal better than we know the geography of the earth. Think of the continent of Africa. In that great country there are mighty tracts, there are vast lakes and ranges of mountains, of which we know but little. We could make a better general map of Africa if it were fastened up on our side of the moon than we actually possess at this moment. There is no spot on the moon so large as an ordinary parish in this country which has not been often observed and measured. There are maps and charts of the moon showing every part of it which is as big as a good-sized field. Indeed, as there are no lunar clouds, the features of its surface are never obscured whenever our own atmosphere will permit us to see through. Artists have frequently sketched the lunar features, and there is plenty of material for them to work on. We have also had photographs taken of the moon, but there is here a difficulty to be encountered which photographers of familiar objects on this earth do not experience. For a photograph to be successful, everybody knows that the first requisite is for the sitter to stay quiet while the plate is being exposed. This is, unhappily, just what the moon cannot do. We endeavour to obviate the difficulty by moving the telescope round so as to follow the moon in its progress. This can be done with considerable accuracy, but, unfortunately, there is another difficulty which lies entirely beyond our control. As the rays of light from the moon perform their journey through hundreds of miles of unsteady air the rays are bent hither and thither, so that the picture which is formed is indistinct. If we are merely *viewing* the moon at the telescope, the quivering, though rather inconvenient, does not prevent us from seeing the object, and we can readily detect the true shape in spite of incessant fluctuations. When, however, these rays fall not on the eye, but on the photographic plate, they produce by their motion a picture which cannot be much magnified without becoming very confused and wanting in sharpness. Hence it happens that for the close study of the moon's appearance we have not up to the present derived much aid from photography.

The adjoining picture (Fig. 38) gives a fair idea of what

Fig. 38.—The Full Moon.

the full moon looks like when viewed through a small telescope. I do not, however, say that the lunar objects can then be observed under favourable conditions; for when the moon is full is the worst time to select for a peep through a telescope. In fact, at this phase you can hardly see anything except slight differences between the colours of different parts. The quarter is the best position for examining the moon ; but even then you can only observe satisfactorily those objects which happen to lie along the border between light and shade. To study the moon properly you must, therefore, watch it during several different phases, from the time when it is the thinnest crescent (just after new moon) until it has again waned to the thinnest crescent (just before the next new moon). We want the relief given by shadows to bring out the full beauty of lunar scenery.

On the map you will first notice the large dark-coloured patches, which are so conspicuous on the moon's face. They are, apparently, the empty basins which great seas once filled. But all the water has now disappeared. These dark parts are, no doubt, a good deal smoother than the rest of the surface; but we can see many little irregularities which tell us that we are not now looking at oceans. The chief features I want you to observe are the curious rings which you see in the figure; there is a very well-marked one a little below the centre, and in the upper part many rings—large and small—are crowded together.

We call them lunar craters. You will see what they are like from the model shown in Fig. 39. But to realise from this picture the proper scale of the object, you should imagine it to be some miles in width. The cliffs which rise all round to form the wall, as well as the mountain which adorns the centre, are quite as high as any of the mountains in Great Britain. You may desire to know how we are able to measure the heights of mountains on the moon? That is what I am now going to show you; and for this purpose we shall look at an imitation lunar crater which I have made with modelling clay. Here is the great ring, or circular enclosure, surrounded by cliffs, and here is a sharp mountain peak rising in the centre. I shall ask to have the beam from the electric lamp turned on our model. You see how prettily it is lighted up. I have placed the lamp so that the beams are sloping; and I have done this with the express object of making the shadows long. In fact, as we look at a lunar crater, which lies on the border

Fig. 39.--Our Model of a Lunar Crater.

between light and shade, the sun illuminates the object under the same conditions as those shown in the figure. I daresay you have often noticed what long shadows are cast at sunset. It is their shadows which teach the astronomer the altitudes of the lunar mountains; for he measures the length of the shadow, and then by a little calculation he can find the height of the object by which that shadow has been cast. I shall suppose that we want to measure the height of a flagstaff (Fig. 40). It is quite possible to do this by merely measuring the length of the shadow which that flagstaff casts at noon. It would not be correct to say that the height of the flagstaff is exactly the length of its shadow. This will, indeed, be the case if you are fortunate enough to make your measurement at London on either the 6th of April or the 5th of September. On all other days in the year a little calculation must be made, which I need not now mention, but which the astronomer, with the aid of his Nautical Almanac, can do in a very few minutes.

Fig. 40.—How we found the Height of the Flagstaff by Measuring the Length of the Shadow.

In a similar manner, by measuring the lengths of the shadows on the moon, and by finding how many miles long these shadows are, we are able to calculate the altitudes of the lunar mountains and of the ranges of cliffs by which the walled plains are surrounded.

ON THE ORIGIN OF THE LUNAR CRATERS.

We have now to offer an explanation of these curious rings, which are the most characteristic features on the moon. To account for them we must look for a moment at some objects on the earth. You have all heard of volcanoes or burning mountains, such as Vesuvius or Etna, which occasionally break out into violent eruptions, and send forth great showers of ashes and torrents of molten lava. In the Sandwich Islands there is a celebrated volcano called Kilauea. It is like a vast lake of lava, so hot that it is actually molten, and glows with heat like red-hot iron. The adventurous tourist who visits this crater can climb to the brink of a lofty range of cliffs which surround it, and gaze down upon the fervid sea beneath. Suppose that by some great change the internal fires which keep this mighty cauldron boiling were to decline and go out, the sea of lava would cease to be liquid, and would ultimately grow hard and cold, and we should then have an immense flat plain, surrounded by a range of cliffs. Elsewhere in the Sandwich Islands examples of extinct craters may be found at the present day. Those who have studied these interesting localities point out how these craters explain the ringed plains in the moon. It seems certain that in ancient days great volcanoes abounded on our satellite, and the rings were often much larger than those on the Sandwich Islands, some of them being one hundred miles or more in diameter. The volcanoes must then have been raging on the moon with a fury altogether unknown in any active volcanoes which this earth can now show. We can also explain the occurrence of the lofty mountain peak which so often rises in the centre of a lunar ring. When the fires had almost subsided, and the floor had grown nearly cold, one last and expiring effort is made by which the congealing surface is burst through at the centre, and a quantity of materials are extruded, which remain as the central mountain to the present day.

I must, however, impress upon you that even the greatest telescopes never exhibit to us any volcanic eruptions at present going on in the moon; in fact, it is most doubtful if any change has been noticed in the features on its surface since the date of the invention of the telescope. The volcanoes sculptured the surface of the moon into the form in which we see it, and that form the moon has preserved for ages, of which we cannot estimate the duration. All the craters and all the volcanoes in the moon can only be described as extinct.

It would be interesting for us to compare the present condition of the volcanoes in the earth, with the present state of the ringed craters in the moon. The noisy volcanoes on our globe are those most talked about; we often hear of Vesuvius being in eruption, and a few years ago (August, 1883) there was a terrific eruption at Krakatoa, during which such a quantity of dust was shot up into the air, that it was borne right round the earth, and produced beautiful sunsets and unwonted sky hues in almost every country in the world. The explosion at Krakatoa made the loudest noise that was ever known. Fortunately such convulsions of the earth do not often happen, for the sea rushed in on the land, and thousands of lives were lost. There are, I believe, at least one hundred volcanoes on different parts of the earth, which are more or less active, but there are many others which have been abandoned by their fires, and which seem to be just as cold and just as extinct as any volcanoes in the moon. Even in our own islands there are abundant remains of ancient volcanoes. Masses of lava are found in many places where now there is no trace of an active volcano. Perhaps there is no more remarkable site in the British Isles than that lofty rock which is crowned by Edinburgh Castle; it is the remnant of a former volcano, while Arthur's Seat, close by, is another. In the centre of France is the beautiful district of Auvergne, in which ancient volcanoes abound; and the lava streams can be traced for miles across the country. These volcanoes have been extinct for thousands of years, during which time the lava has become largely covered with vegetation, and in some places vineyards are cultivated upon it.

We are now able to state the contrast between the earth and the moon, in so far as volcanoes are concerned. On the earth we have some active volcanoes, and a much greater number that are extinct. On the moon we find no active volcanoes, for there all are extinct. I can explain how this difference has arisen, but first let me show you a simple experiment. My assistant will kindly bring to me from that furnace two iron balls, which we placed there before the commencement of this lecture; there they are, you see, both glowing with a bright red heat, for at present they are both equally hot. We will place them on these stands, and allow them to grow cold. One of these balls is a small cannon-ball, four inches in diameter, while the other is only one inch. They are in the same proportion as the earth is to the moon; but look, even while I am speaking the balls have ceased to preserve the same temperature, for the little one has become almost black from loss of its heat, while the large one still looks nearly as red as it did at the beginning; this simple experiment will illustrate the principle that two heated bodies of similar composition will cool at very different rates, if their sizes be different. The small body will always cool faster than the large one. They need not be globes; if you put a poker and a knitting needle into the fire, and leave both there until they are red-hot, and then put them out into the fender, you will speedily find that though they were at the same heat when drawn from the fire, they do not long remain so; indeed, the knitting needle has become cold enough to handle before the poker has ceased to glow. Our experiments were made, no doubt, with small objects, but the law about which they inform us will remain true, even for the greatest objects.

I must ask you to think of events which happened ages ago. Our earth shows many indications of being much hotter within than it is on the surface. The volcanoes themselves are mere outbreaks of incandescent material from inside. Then there are hot springs of water at Bath, which gush out from the earth. There are geysers of hot water in Iceland and in the Yellowstone Park in America, and in other places. And there are other indications also, with which every miner is familiar. Wherever a deep pit is sunk into the earth, the rocks below are always found to be warmer than they are above, and the deeper the pit the greater is the heat that is encountered. Thus, from all over the world we obtain proofs of the present existence of internal heat. Great as the earth is, we must still apply to it the simple common-sense principles that we use in our every-day life here. Let me give an illustration. Suppose that a servant came into the room and placed a jug of water on the table, and that an hour afterwards you went to the jug of water and found it to be perfectly cold, you would not from that fact alone be able to infer anything with certainty, as to whether the water was warm or cold when it was brought in. It might have been perfectly cold, as it is at present though no doubt the water might have been warm at first, and have since cooled down to the temperature of the room during the hour.

Suppose, however, that when you went to the jug of water, which had stood on the table for an hour, you found it tepid, no matter how slightly its temperature might be above that of the room, do you not see the inference you would be able to draw? You would argue in this way: That water has still some heat; it must, of course, be gradually cooling, and therefore it was hotter a minute ago than it is now; it was hotter still two minutes ago, or ten minutes; and must have been very hot and perhaps boiling when it was brought in an hour ago.

I want you to apply exactly the same reasoning to our earth. It is, as I have shown you, still hot and warm inside. Of course, that heat is gradually becoming lost; so that the earth will from year to year gradually cool down, though at an extremely slow rate. But we must look back into what has happened during past ages. Just as we inferred that the jug must have contained very hot water an hour ago from the mere fact that the water in the jug was still warm, so we are entitled to infer from the fact that the earth still contains some heat, that it must ages ago have been exceedingly hot. In fact, the further we look back, the hotter and the hotter do we see the earth growing, until at last we are constrained to think of a period, uncounted ages ago, long ere life began to dawn on this earth, when even the surface of the earth was hot. Back further still we see the earth to have no longer the hard, the dark, and the cold surface we now find; we are to think of it in these primitive ages as a huge glowing mass, in which all the substances that now form the rocks were then incandescent, and even molten material.

There is good reason for knowing that in those early times the moon also melted with heat; and thus our reasoning has conducted us back to a period when there were two great red-hot globes—one of which had four times the diameter of the other—starting on their career of gradually cooling down. Recall our little experiment with the two cooling globes of iron; imagine these globes to preserve their relative proportions, but that one of them was 8,000 miles and the other 2,000 miles across. Ages will, no doubt, elapse ere they part with their heat sufficiently to allow the surfaces to cool and to consolidate. Of this, however, we may be sure, that the small globe will cool the faster, that its outside will become hard sooner than will that of the large one, and that long after the small globe has become cold to the centre, the large one may continue to retain some of its primeval heat. We can thus readily understand why it is that all the volcanoes on the moon have ceased—their day is over. It is over because the moon, being so small, has grown so cold that it no longer sustains the internal fires which are necessary for volcanic outbreaks. Our earth, in consequence of its much greater size, has grown cold more slowly. It has no doubt lost the high temperature on the exterior, and its volcanic energy has probably abated from what it once was. But there is still sufficient power in the subterranean fires to occasionally awaken us by a Krakatoa, or to supply Vesuvius with sufficient materials and vigour for its more frequent outbursts.

The argument shows us that the time will at last come when this earth shall have parted with so large a proportion of its heat that it will be no longer able to provide volcanic phenomena, and then we shall pass into the quiescent stage which the moon attained ages ago.

THE MOVEMENTS OF THE MOON.

Though the moon is going round and round the earth incessantly, yet it always manages to avoid affording us a view of what is on the other side. Our satellite always turns the same face towards us, and though we may reasonably conjecture that the other side is covered like the side we know, with rings and other traces of former volcanoes, yet such knowledge is merely a matter of surmise. In this respect the moon is a very peculiar object. The other great celestial bodies, such as the sun or Jupiter, turn round on their axes, and show us now one side and then the other, with complete impartiality. The way in which the moon revolves may be illustrated by taking your watch and chain, and as you hold the chain at the centre making the watch revolve in a circular path, as shown in Fig. 41. At every point of its path the ring of the watch is, of course, pointed to the centre where the chain is held. If you imagine your eye placed at the centre, the movements of the watch would exemplify the way the moon turns round the earth.

One more point I must explain about the moon before we close this lecture. It is related to a very simple matter. There is nothing more familiar than the fact that a heavy body will fall to the ground. Indeed, it hardly matters what the material of the body may be; for you see I have a small iron ball in one hand and a cork in the other (Fig. 42). I drop them at the same moment, and they reach the ground together. Perhaps you would have expected that the cork would have lagged behind the iron. I try the experiment again and again, and you can see no

Fig. 41.—The Moon always turns the same face to the Earth.

difference in the times of their falling, though I do not say this would be true if they were dropped from the top of the Monument. In general we may say that bodies let drop will fall sixteen feet in the first second.

Wherever we go we find that bodies will always tend to fall in towards the centre of the earth; thus in New

Zealand, at the opposite side of our globe from where we are now standing, bodies will fall up towards us, and this law of falling is obeyed at the top of a mountain as it is down here. No matter how high may be the ascent made in a

Fig. 42.—An Iron Ball and a Cork fall in the same time.

balloon, a body released will fall towards the earth's centre. Of course we can only ascend some five or six miles high, even in the most buoyant of balloons; but we know that the attraction by which bodies are pulled downwards towards the earth extends far beyond this limit. If we could go ten, twenty, or fifty miles up we should still find that the earth tried to pull us down. Nor, even if you could imagine an ascent made to the height of 1,000 miles, would gravitation have ceased. A cork or an iron ball, or any other object dropped from the height of 1,000 miles would assuredly tumble down on the ground below.

Suppose that by some device we were able to soar aloft to a height of 4,000 miles. I name that elevation because we should then be as high above the earth as the centre of the earth is below our feet. We have doubled our distance from the centre of the earth, and the intensity of the gravitation has decreased to one quarter of what it is at the surface. A body which at the earth's surface falls sixteen feet in a second would there fall only four feet in a second, and the apparent weight of any body would be so much reduced that it would seem to weigh only a quarter of what it weighs down here. Thus, the higher and higher we go, the less and less does gravity become; but it does not cease, even at a distance of millions of miles. Therefore you might say that as gravity tries to pull everything down, wherever it may be, why does it not pull down the moon? This is a difficulty which we must carefully consider. Supposing that the earth and the moon were simply held apart, both being at rest, and that then the moon were to be let go, it would no doubt drop down on the earth; but the moon would not do so if at the time it was let go it was thrown sideways; the effect of the earth's pull upon it is then shown in keeping the moon revolving around us instead of allowing it to fly away altogether, as it would have done had the earth not been there to attract it.

We can explain this by an illustration. On the top of a mountain I have placed a big cannon (Fig. 43), and we fire off the cannon, and the bullet flies away in a curved path, with a gradual descent until it falls to the ground. I

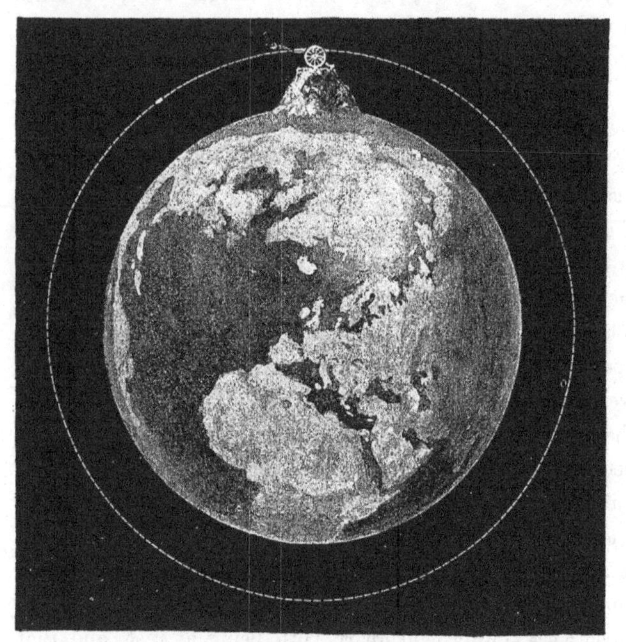

Fig. 43.—An Illustration to explain the Movement of the Moon.

have made the mountain look hundreds of times larger than any mountain could possibly be; and now I want you to imagine a cannon far stronger and gunpowder more potent than any powder or cannon that has ever yet been manufactured. Fire off a bullet with a still greater charge than the last time, and now the path is a much longer one, but still the bullet curves down so as ultimately to fall on the earth. But make now one final shot with a charge sufficiently powerful, and away flies the bullet, following this time the curvature of the earth, so that by the time it has travelled a quarter of the way round it is no nearer to the earth than it was at first, nor has it parted with any of its original speed. Thus, notwithstanding its long journey, the bullet has practically just as much energy as when it first left the muzzle of the cannon. Away it will fly round another quarter of the earth, and still in the same condition it will accomplish the third and the fourth quarters, thus returning to the point from which it started. If we have cleared the cannon out of the way, the bullet will fly again over the mountain-top without having lost any of its speed by its voyage round the earth; therefore it will be in a condition to go round the world permanently. If, then, from the top of a mountain 240,000 miles high a great bullet 2,000 miles in diameter had once been projected with the proper velocity, that bullet would continue for ever to circle round and round the earth, and even though the mountain and the cannon disappeared, the motion would be preserved indefinitely. I do not mean to assert that this is the way in which the moon was actually first given to the earth as an attendant. I have merely used the illustration to show how the continuous revolution of the moon round the earth can exist, notwithstanding that the earth is constantly trying to pull the moon down to its surface.

ON THE POSSIBILITY OF LIFE IN THE MOON.

Astronomers are often asked whether any animals can be living on the moon. The telescope cannot answer that question directly. There are great plains to be seen, but even if there were immense herds of elephants tramping over those plains, our telescopes could not show them. Nor will our instruments pronounce at once whether plants or trees flourish on the moon. The mammoth trees of California are so big that a tunnel has been cut through the trunk of one large enough to give passage for a carriage and pair. Were there trees as big as this on the moon, they would not be visible even from the most famous observatories.

Let us think what we should ourselves experience if we could in some marvellous manner be transferred from the earth to its satellite, and tried to explore that new and wonderful country. Alas! we should find it utterly impossible to live there for an hour, or even for a minute. Troops of difficulties would immediately beset us. The very first would be the want of air. Ponder for a moment on the invariable presence of air around our own globe. Even if you climb to the top of a high mountain, or if you take a lofty voyage in a balloon, you are all the time bathed in air. It is air which supports the balloon, just as a cork is buoyed up by water. Under all circumstances, we must have air to breathe. In that air is oxygen gas, and we must have oxygen incessantly supplied to our lungs to reinvigorate our blood. We require, too, that this oxygen shall be diluted with a much larger amount of nitrogen gas, for our lungs and system of circulation are adapted for abode in that particular mixture of gases which we find here. The atmosphere becomes more and more attenuated the higher we ascend, and apparently terminates altogether some two or three hundred miles over our heads. Beyond the limits of the atmosphere it seems as if empty space would be met with all the way from the earth to the moon. We could not procure a single breath of air, and life would be, of course, impossible. Even at a height of three or four miles, respiration becomes difficult, and doubtless life could not be sustained at all at ten miles high.

It is therefore plain that for a voyage to the moon we should require an ample supply of air, or, at least, of life-giving oxygen, which in some way or other was to be inhaled during the progress of the journey. When at length 240,000 miles had been traversed, and we were about to land on the moon, we would first of all ascertain whether it was surrounded with a coating of air. Most of the globes through space are, so far as we can learn, shrouded and warmed with an enveloping atmosphere of some kind; but, unhappily, the poor moon has been left entirely, or almost entirely, without any such clothing. She is quite bare of atmosphere at all comparable in density or in volume to that which surrounds us, though possibly we do occasionally perceive some traces of air, or of some kind of gas, in small quantities in the lunar valleys.

I am sure each intelligent boy or girl will want to know how we are able to tell all this. We have never been at the moon, and how then can we say that it is nearly destitute of air? Nor can our telescope answer this question immediately, for you could hardly expect to see air, even if it were there. How then can we possibly make such assertions? There are many different ways in which we have learned the absence of air from the moon. I will tell you one of the easiest and the most certain of these methods. First let me say that air is not perfectly transparent. No doubt I can see you, and you can see me, though a good many feet of air lie between us; but when we deal with distances much greater, there is a very simple way in which we can show that air is turbid. In the evening, when the sun is setting and the sky is clear, you can look at him without discomfort; but in the middle of the day you know that it is impossible to look at the sun without shading your eyes with smoked glass or protecting them by some similar contrivance. The reason is, that when the sun is either setting or rising we look at it through an immense thickness of air, which not being perfectly transparent stops some of the light. Thus it is that the sun under these circumstances loses its dazzling brilliancy, and we can view it without discomfort.

At the seaside you can notice the same effect in a different manner. Go out on a fine and clear night, when the stars in their thousands are glittering overhead, and then look down gradually towards the horizon, and you will see the stars becoming fainter and fainter. Indeed, even the brightest star cannot be seen when it is at the horizon, because the atmosphere is not transparent when viewed through an immense thickness.

We can now state the argument by which we may prove that there is little or no air on our satellite. The moon will frequently pass between the earth and a star, and when the star is a really bright one the observations that can be made are of great interest. Let me first describe what we actually see. The star is shining brightly until the moment when the moon eclipses it. Generally speaking, its disappearance is instantaneous. But this would not be the case if the moon were encircled with an atmosphere, for then, as the moon approached the star, the lunar surroundings would come first between the eye and the star. If the moon were coated with air, the light from the star would not be extinguished *instantly;* it would gradually decline, according as it had to pass through more and more of the moon's atmosphere. Thus you would find that the star dwindled down in brightness before the solid body of the moon had advanced far enough to shut it out. The sudden extinction of the star demonstrates the airless state of our satellite.

There would be another insuperable difficulty in adopting the moon as a residence, even supposing that you could get there. Water is absent from its surface. We have examined every part of it, and we find no traces of seas or of oceans, of lakes or rivers; we never see anything like clouds or mists, which are, of course, only water in the vaporous form. We are, therefore, assured that, so far as water is concerned, the moon is an absolute desert. This is, perhaps, the most striking contrast between the aspect of the earth and the aspect of the moon. Were an astronomer on the moon to look at our earth he would find most of its surface shrouded by clouds, and through the openings in these clouds he would see that by far the greater part of this globe was covered by the expanse of ocean; in fact, when the lunar astronomer had realised the prevalence of water upon this earth, either in the form of ocean or cloud, I feel sure he would come to the conclusion that nothing could live here except seals or other amphibious animals.

Owing to the absence of air and water, the moon would be totally disqualified for the support of life of the types in which we know it. For air and water are necessary to every animal, from the humblest animalcule up to whales or elephants. Air and water are necessary for every form of vegetable life, from the lichen which grows on a stone up to the noble old oak that adorns a forest. But even supposing that we could land on the moon, bearing with us an ample supply of oxygen to breathe, and of water to drink, we should find ourselves perplexed and embarrassed, to say the very least of it, by an extraordinary difference that would thrust itself upon our notice. That familiar experience of gravity, or the weights of things, which we have acquired in our residence on a great globe like the earth, would seem ludicrously inappropriate and absurd when we began to walk about on a little globe like the moon. We should be astonished at the transformation by which the weight of everything was much lessened; when you pulled out your watch you would hardly feel it at the end of the chain; it would seem like a mere shell; but yet the watch is all right, it is going as well as ever. Nothing has altered about it except its weight. A big stone attracts your notice, and to your amazement, you find that it does not weigh so much as a piece of wood of the same size would do here. A stone that you could hardly stir on the earth, you can carry about on the moon. Nor is this to be explained by any difference in the constitution of the stone. It will most probably be not very dissimilar to some of the rocks on the earth. The lightness is not in the material, we must seek for some other explanation. Every object on the moon would be found to weigh only one-sixth part of the weight of the same object on the earth. A sturdy labourer at one of the docks can carry one sack of corn on his back here, and he finds that this load is as much as is convenient. He would however discover, were he placed on the moon, that his load had suddenly become lightened to one-sixth part (Fig. 44). The labourer would find that he could carry six sacks of corn on the moon without making a greater effort than the support of a single sack on the earth cost him. To explain how such a change as this

Fig. 44.—The Lessened Gravitation on the Moon.

has occurred, look at these two pictures: one shows the labourer on a small body like the moon, the other shows him on a great globe like the earth. What the labourer feels is not what he thinks he feels. He imagines that it is the weight of the corn, and the corn alone, which produces that pressure on his shoulders which he knows so well. No doubt he is right in a sense, but that is not exactly the way in which the philosopher will look at the same question. What the labourer does actually feel is the attraction between the earth beneath his feet and the corn on his back.

It is this force which produces the pressure on his shoulders. Its magnitude no doubt depends upon the quantity of corn in the sack, but it also depends on the quantity of matter on the earth beneath his feet. In fact, the force between two attracting bodies depends upon the masses or weights of both the attracting bodies. When the labourer is transferred to the moon, of which the mass is so much less than that of the earth, the attraction is less there than it is here, even though the corn is the same in the two cases.

Many odd instances could be given of the extraordinary consequences of life on a world where all weights are reduced to a sixth part. One occurred to me the other day when I saw a postman going his rounds with an amazing load of Christmas presents and parcels. I thought, how much happier must be the lot of a postman on the moon, if such functionaries are wanted there! All the presents of toys or more substantial donations might be the same as before, the only alteration would be that they would not have felt nearly so heavy. A box which contains a pound of chocolate bonbons might still contain exactly the same quantity of sweetmeat on the moon, but the exertion of carrying it would be reduced to one-sixth. It would only weigh as much as two or three ounces do on the earth. Our streets provide another admirable illustration of the drawbacks of our life here as compared with the facilities offered by life on the moon. I feel quite confident that no perambulators can be necessary there. I cannot indeed say that there are babies to be found on the moon, but of this I am certain, that even if the lunar babies were as plump and as sturdy as ours, they must still only weigh about a sixth as much as ours do. A lunar nurse would scorn to use a perambulator, even for a pair of twins; she might take them both out on her arm for an airing, and even then only bear one-third of the load that her terrestrial sister must sustain if she is carrying but a single child.

The lightness of bodies in the moon would entirely transform many of our most familiar games. In cricket, for instance, I don't think the bowling would be so much affected, but the hits on the moon would be truly terrific. I believe an exceptionally good throw of the cricket-ball here is about a hundred yards, but the same man using the same ball and giving the same run to it, would send the ball six hundred yards on the moon. So, too, every hit would in the lunar game carry the ball to six times the distance it does here. Football would show a striking development in lunar play; a good kick would not only send the ball over the cross-bar, but it would go soaring over the houses, and perhaps drop in the next parish.

Our own bodies would, of course, participate in the general buoyancy, so that while muscular power remained unabated, we should be almost able to run and jump as if we had on the famous seven-league boots. I have seen an athlete in a circus jump over ten horses placed side by side. The same athlete, making the same effort, would jump over sixty horses on the moon.

A run with a pack of lunar foxhounds would indeed be a marvellous spectacle. There need be no looking round by timid horsemen to find open roads or easy gaps. The five-barred gate itself would be utterly despised by a huntsman who could easily clear a hay-rick. Nor would the farmer be astonished if all the field jumped over his house without disturbing a slate on the roof. It would hardly be worth taking a serious jump to clear a canal unless there was a road and a railway or so, which could be disposed of at the same time.

To illustrate this subject of gravitation in another way, suppose that we were to be transferred from this earth to some globe much greater than the earth—to a globe, for instance, as large and massive as the sun. We can then show that the weight of every object would be increased. Indeed, everything would weigh about twenty-seven times as much as we find it does here. To pull out your watch would be to hoist a weight of about five or six pounds out of your pocket, but I do not see how you could do it, for even to raise your arm would be impossible; it would feel heavier by far than if it were made of solid lead. It is, perhaps, conceivable that you might stand upright for a moment, particularly if you had a wall to lean up against; but of this I feel certain, that if you once got down on the ground, it would be utterly out of your power to rise again.

These illustrations will at least answer one purpose: they will show how difficult it is for us to form any opinion as to the presence or of the absence of life on the other globes in space. We are just adapted in every way for a residence on this particular earth of a particular size and climate, and with atmosphere of a particular composition. Within certain slender limits our vital powers can become accommodated to change, but the conditions of other worlds seem to be so utterly different from those we find here, that it would probably be quite impossible for beings constituted as we are to remain alive for five minutes on any other globe in space.

It is, however, quite another question as to whether there

may not be inhabitants of some kind on many of the other splendid globes. We have through the wide extent of space inconceivable myriads of worlds, presenting, no doubt, every variety of size and climate, of atmosphere and soil. It seems quite preposterous to imagine that from among all these globes ours alone should be the abode of life. The most reasonable conclusion for us to come to is that these bodies may be endowed with life of types which are just as appropriate to the physical conditions around them as is the life, both animal and vegetable, on this globe to the special circumstances in which it is placed.

LECTURE III.

THE INNER PLANETS.

Mercury, Venus, and Mars—How to Make a Drawing of our System—The Planet Mercury—The Planet Venus—The Transit of Venus—Venus as a World—The Planet Mars and his Movements—The Ellipse—The Discoveries made by Tycho and Kepler—The Discoveries made by Newton—The Geography of Mars—The Satellites of Mars—How the Telescope aids in viewing Faint Objects—The Asteroids, or Small Planets.

MERCURY, VENUS, AND MARS.

WE can hardly think of either the sun or the moon as a world in the sense in which our earth is a world, but there are some bodies called planets which seem more like worlds, and it is about them that we are now going to talk. Besides our Earth there are seven planets of considerable size, and a whole host of insignificant little ones. These planets are like ours in a good many respects. One of them, Venus, is about the same size as this earth ; but the two others, Mercury and Mars, are a good deal smaller. There are also some planets very much larger than any of these, namely, Jupiter, Saturn, Uranus, and Neptune. We shall in this lecture chiefly discuss three bodies, namely, Mercury, Venus, and Mars, which, with the earth, form the group of "inner" planets.

The planets are all members of the great family dependent on the sun. Venus and the earth may be considered the pair of twins, alike in size and in the duration of their day. Mercury and Mars are the babies of the system. The big brothers are Jupiter and Saturn. All the planets revolve around the sun, and derive their light and their heat from his beams. We should like to get a little closer to some of our fellow-planets, and learn their actual geography. Unfortunately, even under the most favourable circumstances, they are a very long way off. They are many millions of miles distant, and are always at least a hundred times as far as the moon. But far as the planets may be, astronomers are familiar with their existence, and so have they been for ages past. I can give you a curious proof of this. You remember how we said the first and the second days of the week were called after the sun and the moon, Sun-day and Moon-day, or Monday, respectively. Let us see about the other days. Tuesday is not quite so obvious, but translate it into French and we have at once *Mardi;* this word means nothing but Mars' day, and our Tuesday means exactly the same. Wednesday is also readily interpreted by the French word *Mercredi,* or Mercury's day, while Venus corresponds to Friday. Jupiter's day is Thursday, while Saturn's day is naturally Saturday. The familiar names of the days of the week are thus associated with the seven moving celestial bodies which have been known for uncounted ages.

HOW TO MAKE A DRAWING OF OUR SYSTEM.

I want everyone who reads this book to make a little drawing of the sun and the planets. The apparatus that you will need is a pair of compasses; any sort of compasses that will carry a bit of pencil will do. You must also get a little scale that has inches and parts of inches divided upon it; any carpenter's rule will answer. The drawing is intended to give a notion of the true sizes and positions of the fine family of which the earth is one member. The figure I have given (Fig. 45) is not on so large a scale as

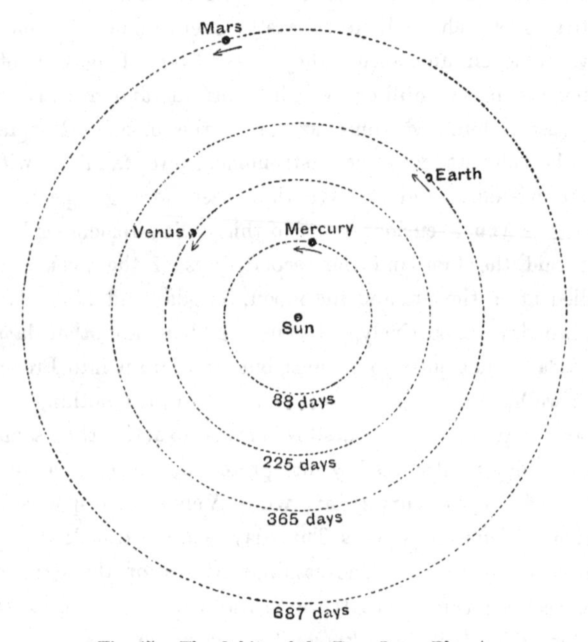

Fig. 45.—The Orbits of the Four Inner Planets.

that which I ask you to use, and which I shall here mention. Try and do the work neatly, and then pin up your little drawings where you will be able to see them every day until you are quite familiar with the notion of what we mean by our solar system.

First open the compasses one inch, and then describe a

circle, and mark a dot on this as "MERCURY, in neat letters, and also write on the circle "88 days." At the centre you are to show the "SUN." This circle gives the track followed by Mercury in its journey round the sun in the period of 88 days. Next open your compasses to $1\frac{3}{4}$ in., which you must do accurately by the scale. The circle drawn with this radius shows the relative size of the path of Venus, and to indicate the periodic time, you should mark it, "225 days." The next circle you have to draw is a very interesting one. The compass is to be opened $2\frac{1}{2}$ in. this time, and the path that it makes is to be marked "365 days." This shows the high road along which we ourselves journey every year, along which we are, indeed, journeying at this moment. If you wanted to obtain from your figure any notions of the true dimensions of the system, the path of the earth will be the most convenient means of doing so. The earth is 93,000,000 miles from the sun, and our drawing shows its orbit as a circle of $2\frac{1}{2}$ in. radius. It follows that each inch on our little scale will correspond to about 37,000,000 miles. As, therefore, the radius of the orbit of Mercury has been taken to be one inch, it follows that the distance of Mercury from the sun is about 37,000,000 miles.

We have, however, still one more circle to draw before we complete this little sketch. The compass must now open to four inches, and a circle which represents the orbit of Mars is then to be drawn. We mark on this "687 days," and the inner part of the solar system is then fully represented. You see this diagram shows how our earth is in every sense a planet. It happens that one of the four planets revolves outside the earth's path, while there are two inside. By marking the days on the circles which show the periods of the planets, you perceive that the further a planet is from the sun, the longer is the time that it takes to go round. Perhaps you will not be surprised at this, for the length of the journey is, of course, greater in the greater orbits; but this consideration will not entirely explain the augmentation of the time of revolution. The further a planet is from the sun, the more slowly does it actually move, and therefore, for a double reason, the larger orbit will take a longer time. From London to Brighton is a much longer journey than from London to Greenwich, and, therefore, the journey by rail to Brighton will, of course, be a longer one than by rail to Greenwich. But suppose that you compared the railway journey to Greenwich with the journey, not by rail, but by coach, to Brighton, here the comparative slowness of the coach would form another reason besides the greater length of the journey for making the Brighton trip a much more tedious one than that to Greenwich. Mars may be likened to the coach, which has to go all the way to Brighton, while Mercury may be likened to the train which flies along over the very short journey to Greenwich.

We can easily show from our little sketch that Mercury must be moving more quickly than Mars, for the radii of the two circles are respectively one inch and four inches, therefore the circumference of Mars' path must be four times that representing the orbit of Mercury. If Mars moved as fast as Mercury, he would, of course, require only four times as many days to complete his large path as Mercury takes for his small path; but four times 88 is 352, and, consequently, Mars ought to get round in 352 days if he moved as fast as

Mercury does. As a matter of fact, Mars requires nearly twice that number of days; indeed, no less than 687, and hence we infer that the average speed of Mars cannot be much more than half that of Mercury.

To duly appreciate the position of the earth with regard to its brothers and sisters in the sun's family, it will be necessary to use your compasses in drawing another little sketch, by which the sizes of the four bodies themselves shall be fairly represented. Remember that the last drawing showed nothing about the sizes of the bodies, it merely exhibited the dimensions of the paths in which they moved. As Mercury is the smallest globe of the four we shall open the compasses half an inch, and describe a circle to represent it. The earth and Venus are so nearly the same size (though the earth is a trifle the larger) that it is not even necessary to try and show the difference between them, so we shall represent both bodies by circles, each $1\frac{1}{4}$ inches in radius. Mars, like Mercury, is one of the globes smaller than the earth, and the circle for it will have a radius of $\frac{3}{4}$ of an inch. You should draw these figures neatly, and by a little shading make them look like globes. It would be better still if you were to make actual models, taking care, of course, to give each of them the exact size. A comparative view of the principal planets is shown in Fig. 46.

THE PLANET MERCURY.

Quicksilver is a bright and pretty metal, and, unlike every other metal, it is a liquid. If you mix it with some tin foil, it forms a sort of paste, which when put on the back of a piece of glass makes a looking-glass. If you

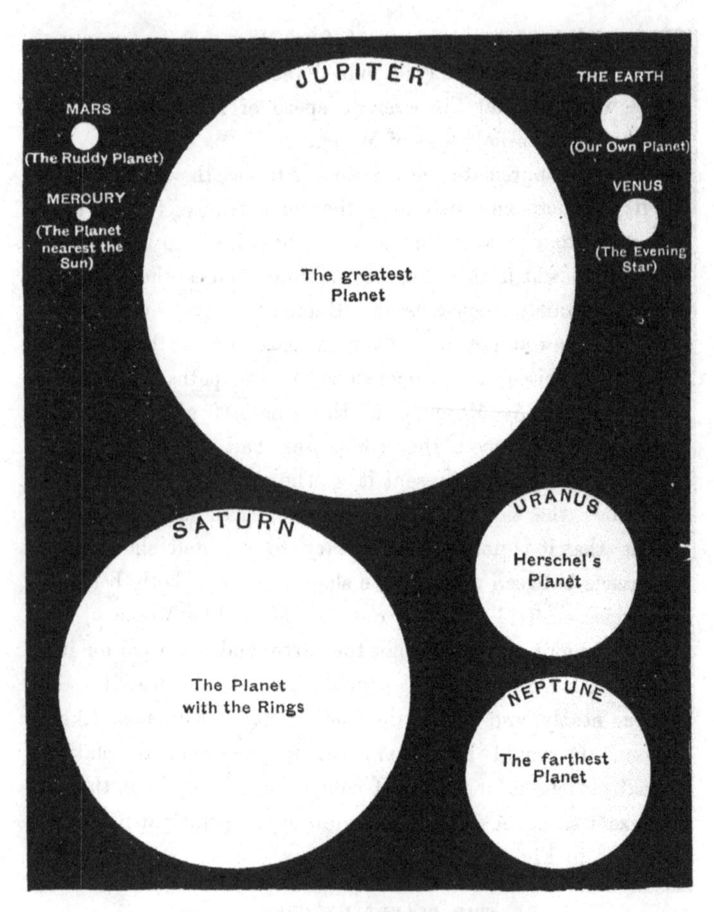

Fig. 46.—Comparative Sizes of the Planets.

spill quicksilver, it is a difficult task to gather the liquid up again. It breaks into little drops, and you cannot easily lift them with your fingers; they slip away and elude you.

Quicksilver will run easily through a hole so small that water would hardly pass, and it is so heavy that an iron nail or a bunch of keys will float upon it. Now, this heavy, bright, nimble metal is known by another name besides quicksilver: a chemist would call it Mercury, and the astronomers use exactly the same word to denote a pretty, bright, nimble, and heavy planet which seems to try to elude our vision. Though Mercury is so hard to see, yet it was discovered so long ago, that all record is lost of who the discoverer was. I do not, however, assert that either the planet has been named after the metal, or the metal after the planet, although we use the same word for both.

Fig. 47.

You must take special pains if you want to see the planet Mercury, for during the greater part of the year it is not to be seen at all. Every now and then a glimpse is to be had, but you must be on the alert to look out just after sunset, or you must be up very early in the morning so as to see it just before sunrise. Mercury is always found to be in attendance on the sun, so that you must search for him near the sun, that is low down in the west in the evenings, or low down in the east in the mornings.

Mercury revolves in a path which lies inside that of Venus, and it is therefore nearer to the sun. Indeed, Mercury is generally so close to the sun that it cannot be seen at all. Like every other planet, Mercury is lighted by the sun's rays, and shows phases in the telescope just as the moon does (Fig. 47). In this figure the different apparent sizes of the planet at different parts of its path are shown. Of course the nearer Mercury is to the earth the larger does it seem.

If we can only see Mercury so rarely, and if even then it is a very long way off, does it not seem strange that we can tell how heavy it is? Even if we had a big enough scales to hold a planet and the necessary weights, what, it may be asked, would be the use of the scales when the body to be weighed was about a hundred millions of miles away? Of course the weighing of a planet must be conducted in some manner totally different from the kind of weighing that we ordinarily use. Astronomers have, however, various methods for weighing these big globes, even though they can never touch them. We do not, of course, want to know how many pounds, or how many millions of tons they contain, there is no use in trying to express the weight in that way. It gives no conception of a planet's true importance. One world must be compared with another world, and we therefore estimate the weights of the other worlds by comparing them with that of our own. We accordingly have to consider Mercury placed beside the earth, and to see which of the two bodies is the bigger and the heavier, or what is the proportion between them. It so happens that Mercury, viewed as a world, is a very small body. It is a good deal less in size than our earth, and it is not nearly so massive. To show you how we found this out, I shall venture on a little story. It will explain one of the strange devices that astronomers have to use when they want to weigh a distant body in space.

There was once, and there is still, a little comet which flits about the sky; we shall call it after the name of its discoverer, Encke. There are sometimes splendid comets which everybody can see—we will talk about these afterwards—but Encke is not such a one. It is very faint and delicate, but astronomers are interested in it, and they always look out for it with their telescopes; indeed, they could not see the poor little thing without them. Encke goes for long journeys through space—so far that it becomes quite invisible, and remains out of sight for two or three years. All this time it is tearing along at a tremendous speed. If you were to take a ride on the comet, it would whirl you along far swifter than if you were sitting on a cannon-ball. When the comet has reached the end of its journey, then it wheels round and returns by a different road until at last it comes near enough to show itself. Astronomers give it all the welcome they can, but it won't remain, sometimes it will hardly stay long enough for us to observe that it has come at all, and sometimes it is so thin and worn after all its wanderings that we are hardly able to see it. Poor Encke never takes any rest; even during its brief visit to us it is scampering along all the time, and then again it darts off, gradually to sink into the blue, whither even our best telescopes cannot follow it. No more is there to be seen of Encke for another three years, when again it will come back for awhile. Encke is like the cuckoo, which only comes for a brief visit every spring, and even then is often not heard by many who dearly love his welcome note; but Encke is a greater stranger than the cuckoo, for the comet never repeats his visit of a few weeks more than once in three years; and he is then so shy that usually very few catch a glimpse of him.

An astronomer and a mathematician were great friends, and they used to help each other in their work. The astronomer watched Encke's comet, noted exactly where it was, on each night it was visible, and then told the mathematician all he had seen. Provided with this information the mathematician sharpens his pencils, sits down at his desk, and begins to work long columns of figures, until at length he discovers how to make a time table which shall set forth the wanderings of Encke. He is able to verify the accuracy of his table in a very unmistakable way by venturing upon prophecies. The mathematician predicts to the astronomer the very day and the very hour at which the comet will reappear. He even indicates the very part of the heavens to which the telescope must be directed, in order to greet the wanderer on its return. When the time comes the astronomer finds that his friend has been a true prophet; there is the comet on the expected day, and in the expected constellation.

This happens again and again, so that the mathematician, with his pencil and his figures, marks stage by stage the progress of Encke through the years of his invisible voyage. At each moment he knows where the comet is situated, though utterly unable to see it.

The joint labours of the two friends having thus discovered law and order in the movements of the comet, you may judge of their dismay when on one occasion Encke disappointed them. He appeared, it is true, but then he was a little late, and he was also not in the spot where he was expected. There was nearly being a serious difference between the two friends. The astronomer accused the mathematician of having made mistakes in his figures, the mathematician retorted that the astronomer must have made some blunder in his observations. A quarrel was imminent, when at last it was suggested to interrogate Encke himself, and see whether he could offer any explanation. The mathematician employed peculiar methods that I could not explain, so I shall transform his processes into a dialogue between the mathematician and the offending comet.

"You are late," said the mathematician to the comet. "You have not turned up at the time I expected you, nor are you exactly in the right place; nor, indeed, for that matter, are you now moving exactly as you ought to do. In fact, you are entirely out of order, and what explanation have you to give of this irregularity?"

You see the mathematician felt quite confident that there must have been some cause at work that he did not know of. Mathematicians have one great privilege: they are the only people in the world that never make any mistakes. If the mathematician knew accurately all the various influences that were at work on the comet, he could, by his figures, have found out exactly where the comet would be. If the comet was not there, it is inevitable that there must have been something or other acting upon the comet, of which the mathematician was in ignorance.

The comet, like every other transgressor, immediately began to make excuses, and to shuffle off the blame on somebody else. "I was," said Encke, "going quietly on my rounds as usual. I was following out stage by stage the track that you know so well, and I would certainly have completed my journey and have arrived here in good time and in the spot where you expected me had I been let alone, but unfortunately I was not let alone. In the course of my long travels—but at a time when you could not have seen me—I had the misfortune to come very close to a little planet, of which I daresay you have heard—it is called Mercury. I did not want to interfere with Mercury; I was only anxious to hurry past and keep on my journey, but he was meddlesome, and began to interfere with me, and I had a great deal of trouble to get free from him, but at last I did shake him off. I kept my pace as well as I could afterwards, but I could not make up the lost time, and consequently I am here a little late. I know I am not just where I ought to be, nor am I now moving quite as you expect me to do; the fact is, I have not yet quite recovered from the bad treatment I have experienced."

The astronomer and the mathematician proceeded to test this story. They found out what Mercury was doing; they knew where he was at the time, and they ascertained that what the comet had said was true, and that it had come very close indeed to the planet. The astronomer was quite satisfied, and was proposing to turn to some other matter, when the mathematician said—

"Tarry a moment, my friend. It is the part of a wise man to extract special benefit from mishaps and disasters. Let us see whether the tribulations of poor Encke cannot be made to afford some very valuable information. We expected to find Encke here. Well, he is not here—he is there, a little way off. Let us measure the distance between the place where Encke is, and the place where he ought to have been."

This the astronomer did. "Well," he said, "what will this tell you? It merely expresses the amount of delinquency on the part of Encke."

"No doubt," said the mathematician, "that is so; but we must remember that the delinquency, as you call it, was caused by Mercury. The bigger and the heavier was Mercury, the greater would be his power of doing mischief, the more would he have troubled poor Encke, and the larger would be the derangement of the comet in consequence of the unfortunate incident. We have measured how much Encke has actually been led astray. Had Mercury been heavier than he is, that quantity would have been larger; and if Mercury had been lighter than he is, you would not, of course, have found so large an error in the comet."

We may illustrate what is meant in this way. A steamer sails from Liverpool to New York, and under favourable circumstances the voyage across the Atlantic should be accomplished within a week. But supposing that in the middle of the ocean a storm is encountered, by which the ship is driven from her course. She will, of course, be delayed, and her voyage will be lengthened. A trifling storm, perhaps, she will not mind, but a heavy storm might delay her six hours, a still greater storm might keep her back half a day, while cases are not unfrequent in which the delay has amounted to one day, or two days, or even more.

The delay which the ship has experienced may be taken as a measure of the vehemence of the storm. I am not supposing that her machinery has broken down; of course, that sometimes happens at sea, as do calamities of a far more tragic nature. I am merely supposing the ship to be exposed to very heavy weather, from which she emerges just as sound as she was when the storm began. In such cases as this we may reasonably measure the intensity of the storm by the number of hours' delay to which the passengers were subjected. "The weather we had was much worse than the weather you had," one traveller may say to another. "Our ship was two days late, while you escaped with a loss of one day."

When the comet at last sailed back into port after a cruise of three years through space, the number of hours by which it was late expressed the vehemence of the storm it experienced. The only storm that the comet would have met with, at least in so far as our present object is concerned, was the breeze that it had with Mercury. The weight of Mercury was, therefore, involved in the delay of the comet. In fact, the delay was a measure of the weight of the planet. I do not attempt to describe to you all the long work through which the mathematician had to plod before he could ascertain the weight of Mercury. It was a very long and a very hard sum, but at last his calculations arrived at the answer, and showed that Mercury must be a light globe compared to the earth. In fact, it would take twenty-five globes, each equal to Mercury, to weigh as much as the earth.

I daresay you will think that this was a very long and roundabout way of weighing. I don't recommend it for ordinary purposes. The grocer would find Encke rather an impracticable method of finding the weight of a pound

Fig. 48.

of sugar. Supposing, however, we had to weigh a mountain, or rather a body, which was bigger than fifty thousand mountains, and which was also many millions of miles away, all sorts of expedients would have to be resorted to. I have told you one of them. If you feel any doubts as to the accuracy with which such weighings can be made, then I must tell you that there are many other methods, and that these all give the same result.

As to what the globe of Mercury may be like, we hardly know anything. We can see little or nothing of the nature of its surface. We only perceive the planet to be a ball, brightly lighted by the sun, and we can hardly discern features thereon, as we are able to do on many of the other planets.

THE PLANET VENUS.

You will have no difficulty in recognising Venus, but you must choose the right time to look out for her. In the first place, you need never expect to see Venus very late at night. You should look for the planet in the evening, as soon as it is dark, towards the west, or in the morning, a little before sunrise, towards the east. I do not, however, say that you can always see Venus, either before sunrise or after sunset. In fact, for a large part of the year Venus is not to be seen at all. You should therefore consult the almanac, and unless you find that Venus is stated to be an evening star or a morning star, you need not trouble to search for it. I may, however, tell you that Venus can never be an evening star and a morning star at the same time. If you are watching it this evening after sundown, there is no use in getting up early in the morning to look out for it again. The planet will remain for several weeks a splendid object after sunset, and then will gradually disappear from the west, and in a couple of months later will be the morning star in the East. Venus requires, altogether, a year and seven months to run through her changes, so that if you find her a bright evening star to-night, you may feel sure that she was a bright evening star a year and seven months ago, and that she will be a bright evening star in a year and seven months to come. Nor must you ever expect to see Venus right over-head ; she is always to the west or to the east.

The splendour of Venus, when she is at her best, will prevent you from mistaking this planet for an ordinary star. She is then more than twenty times as bright as any star in the heavens. In town you will sometimes catch a beautiful view of Venus when, between the lights of the city and the obscurity which generally covers it, hardly a star, properly so called, can be seen. The most conclusive proof of the unrivalled brightness of Venus is found in the fact that she can be recognised in broad daylight without a telescope. Even on the brightest of June days the lovely planet is sometimes to be discerned like a morsel of white cloud on the perfect azure of the sky.

Venus is so brilliant that perhaps you will hardly credit me when I tell you that she has no more light of her own than has a stone or a handful of earth, or a button. How can this be, you will say, for as we see the planet so exquisitely beautiful, how can she be merely a huge stone hung up in the heavens ? The fact is that Venus shines by light not her own, but by light which falls upon her from the sun. She is lighted up just as the moon, or just as our own earth is lighted. Her radiance merely arises from the sunbeams which fall upon her. It seems at first surprising that mere sunbeams on the planet can give her the brilliancy that is sometimes so attractive. Let me show you an illustration which will, I trust, convince you that sunbeams will be adequate even for the glory of Venus.

Here is a button. I hang it by a piece of fine thread, and when I dip it into the beam from the electric lamp, look at the brilliancy with which the mimic planet glitters. You do not see the shape of the button, it is too small for that, you merely see it as a brilliant gem, radiating light all around. Therefore, we need not be surprised to learn that the brilliancy of the evening star is borrowed from the sun, and that, if while we are looking at the planet in the evening, the sun were to be suddenly extinguished, the planet would also vanish from view, though the stars would shine as before.

Thus we explain the appearance of Venus. The evening star is a beautiful, luminous point, but no shape can be discerned with the unaided eye. When, however, the telescope is turned towards Venus we have the delightful spectacle of a tiny moon, which goes through its phases just like our own moon. When first seen Venus will often be like the moon at the quarter, and then it will pass to the crescent shape. Then the crescent will get thinner and thinner, and next will follow a brief period of invisibility before the appearance of Venus as the morning star. It seems at first a little strange that Venus when brightest should not be full like the moon, which under similar circumstances is, of course, a complete circle of light. The planet, however, has a very marked crescent-shaped form under these circumstances. But at this time the planet is so near us that the gain of brilliancy from the diminution of distance more than compensates for the small part of the illuminated side which is turned towards us.

You ought all to try and get someone to show you Venus in a telescope. A very large instrument is not necessary, and I feel sure you will be delighted to see the beautiful moon-shaped planet. You will then have no difficulty in understanding how the beauty and brightness have come from the sun. The changes in the crescent merely depend upon the proportion of the illuminated side which is turned towards us. Were Venus itself a sunlike body we should, of course, see no crescent, but only a bright circle of light.

Fig. 49.—To show that Venus shines by Sunlight.

In Fig. 49 you will see a young astronomer surveying Venus with a telescope. I have not attempted to show things in their proper proportions—that would be impossible. The sun has set, so that his beams do not reach the astronomer. Night has begun where he is; but the sunbeams fall on Venus, and light her up on that side turned towards the sun. A part of this lighted side is, of course, seen by the telescope which the astronomer is using, and thus the planet seems to him like a crescent of light.

THE TRANSIT OF VENUS.

One might naturally think from Fig. 45 that Venus must pass at every revolution directly between the earth and the sun; and therefore it might appear that what is called the transit of Venus across the sun ought to occur every time between the appearance of the planet as the evening star and the next following appearance as the morning star. No doubt on each of these occasions Venus seems to closely approach the sun; but the two orbits do not lie quite in the same plane, and hence the planet usually passes just over or just under the sun, so that it is a very rare event indeed for her to come right in front of the sun. But this does sometimes happen. It happened, for instance, in the year 1874, and again in the year 1882; but, alas! I cannot hold out to you the prospect of ever seeing a repetition. There will be no further occurrence of the transit of Venus until the year 2004, though there will be another eight years later, in the year 2012.

It seems rather odd that the transits of Venus should occur in a pair at an interval of eight years, and that then a period of much more than a century should have to elapse before there can be a repetition of a pair. This is in consequence of a curious relation between the motion of Venus and the motion of the earth, which I must endeavour to explain with the help of a little illustration.

Let us suppose a clock with ordinary numbers round the dial, but so arranged that the slowly-moving short hand requires 365·26 days to complete one revolution round the dial, while the more rapidly-moving long hand revolves in 224·70 days. The short hand will then go round once in a year, and the long hand once during the revolution of Venus. Let us suppose that both hands start together from XII, then in 224·70 days the long hand is round to XII again, but the short hand will have only advanced to about VII, and by the time it reaches XII the long hand will have completed a large part of a second circuit. It happens that the two numbers 224·70 and 365·26 are very nearly in the ratio of 8 to 13. In fact, if the numbers had only been 224·8 and 365·3 respectively, they would be exactly in the ratio of 8 to 13. It, therefore, follows that eight revolutions of the short hand must occupy very nearly the same time as thirteen revolutions of the long hand. After eight years the short hand will of course be found again at XII; and at the same moment the long hand will also be back almost exactly at XII, after completing thirteen revolutions.

We can now understand why the transits, when they do occur, generally arrive in pairs at an interval of eight years. Suppose that at a certain time Venus happens to interpose directly between earth and sun, then, when eight years have elapsed, the earth is, of course, restored for the eighth time since the first transit to the same place, and Venus has returned to almost the same spot for the thirteenth time. The two bodies are practically in the same condition as they were at first, and, therefore, Venus again intervenes, and the planet is beheld as a black spot on the sun's surface. We must not push this argument too far; the relation between the two periods of revolution, though nearly, is not exactly 8 to 13.

The consequence is that when yet another eight years has come round, the planet passes a little above the sun or a little below the sun, and thus a third occurrence of the transit is prevented until more than a century has elapsed. The transit will then take place at the opposite side of the path.

We were fortunate enough to be able to see the transit

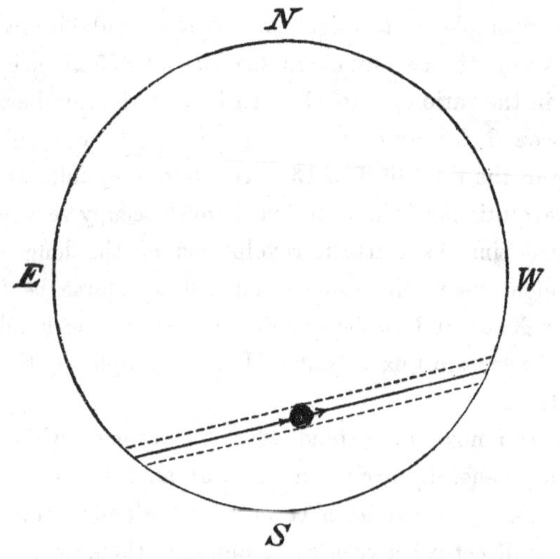

Fig. 50.—Venus in transit across the Sun.

of Venus in 1882 from Great Britain. Perhaps I should say a part of the transit, for the sun had set long before the planet had finished its journey across the disc. Venus looked like a small round black spot, stealing in on the bright surface of the sun and gradually advancing along the short chord that formed its track.

An immense deal of trouble was taken in 1882, as well as in 1874, to observe this rare occurrence. Expeditions were sent to various places over the earth where the circumstances were favourable. Indeed, I do not suppose that there was ever any other celestial event about which so much interest was created. You may inquire why it was that astronomers were so anxious to see Venus in this unusual manner. The true reason why the event attracted so much attention was not solely on account of its beauty or its singularity: it was because the transit of Venus affords us a valuable means of learning the distance of the sun. When observations of the transit of Venus made at opposite sides of the earth are brought together, we are enabled to calculate from them the distance of Venus, and knowing that, we can find the distance of the sun and the distances and the sizes of the planets. This is very valuable information; but you would have to read some rather hard books on astronomy if you wanted to understand clearly how it is that the transit of Venus tells us all these wonderful things. I may, however, say that the principle of the method is really the same as that mentioned on pp. 18—24. When you remember that not we ourselves, nor our children, and hardly our grandchildren, will ever be able to see another transit of Venus, you will, perhaps, not be surprised that we tried to make the most of such transits as we have witnessed.

VENUS AS A WORLD.

Though Venus exhibits such pretty crescents in the telescope, yet I must say that in other respects a view of the planet is rather disappointing. Venus is adorned by such a very bright dress of sunbeams that we can see but little more than those sunbeams, and we can hardly make out anything of the actual nature of the planet itself. We can sometimes discern faint marks upon the globe, but it is impossible even to make a conjecture of what the Venus country is like. This is greatly to be regretted, for Venus approaches comparatively close to the earth, and is a world so like our own in size and other circumstances that we feel a legitimate curiosity to learn something more about her. Indeed the features on Venus seem so vague and ill-defined that we cannot rely on them sufficiently to pronounce with any certainty on the interesting question as to the time that Venus takes to rotate on her axis. Venus is one of the few globes which might conceivably be the abode of beings not very widely different from ourselves. In one condition especially—namely, that of weight—she resembles the earth so closely that those bodies in which we are here clothed would probably be adapted, so far as strength is concerned, for a residence on the sister planet. Our muscles would not be unnecessarily strong, as they would be on the moon, nor should we find them too weak, as they would certainly prove to be were we placed on one of the very heavy bodies of our system. Nor need the temperature of Venus be regarded as presenting any insuperable difficulties. It is, of course, nearer to the sun than we are, but then climate depends on other conditions besides nearness to the sun, so that the question as to whether Venus would be too hot for our abode could not be readily decided. The composition of the atmosphere surrounding the planet would be the most material point in deciding whether terrestrial beings could live there. I think it in the highest degree unlikely that the atmosphere of Venus should chance to suit us in the requisite particulars, and, therefore, I think there is not much likelihood that Venus is inhabited by any men, women, or children resembling those on this earth.

THE PLANET MARS AND HIS MOVEMENTS.

The path of the earth lies between the contiguous orbits of the planets Venus and Mars. It is natural for us to endeavour to learn what we can about our neighbours. We ought to know something, at all events, as to the people who live next door to us on each side. I have, however, already said that we cannot observe much upon Venus, except her brilliance. The case is very different with respect to Mars. He is a planet which we are fortunately enabled to study minutely, and he is full of interest when we examine him through a good telescope.

The right season for observing Mars must, of course, be awaited, as he is not always visible. Such seasons recur about every two years, and then for months together Mars will be a brilliant object in the skies every night. Nor has Mars necessarily to be sought in the early morn or immediately after sunset in the manner we have already described for Venus and Mercury. At the time Mars is at his best he comes into the highest position at midnight, and he can generally be seen for hours before, and be followed for hours subsequently. You may, however, find some difficulty in recognising him. You probably would not at first be able to distinguish Mars from a fixed star. No doubt this planet is a ruddy object, but some stars are also ruddy, and this is at the best a very insecure characteristic for identification. I cannot give you any more general directions, except that you should get your papa to point out Mars to you the next time it is visible. It is just conceivable that papa himself might not know how to find Mars. If so, the sooner he gets a set of star maps and begins to teach himself and to teach you, the better it will be for you both.

Mars, though apparently so like a star, differs in some essential points from any star in the sky. The stars proper are all fixed in the constellations, and they never change their mutual positions. The groups which form the Great Bear or the Belt of Orion do not alter, they are just the same now as they were centuries ago. But the case is very different with a planet such as Mars. The very word planet means a *wanderer*, and it is justly applied, because Mars, instead of staying permanently in any one constellation, goes constantly roaming about from one group to the other. He is a very restless spirit: sometimes he pays his respects to the heavenly Twins, and is found near Castor and Pollux in Gemini, then he goes off and has a brief sojourn with the Bull, but it looks as if that fierce animal got tired of his company and hunted him off to the Lion. His quarters then become still more critical. Sometimes it looks as if he desired to seek for peace beneath the waters, and so he visits Aquarius, while at other times he is found in dangerous proximity to the Crab's claws.

Mars cannot even make up his mind to run steadily round the heavens in one direction: sometimes he will bolt off rapidly, then pause for a while, and turn back again; then the original impulse will return, and he will resume his journey in the direction he at first intended. It is no wonder that I am not able to give you very explicit directions as to how you are to secure a sight of a truant whose movements are so erratic. Yet there is a certain order underlying all his movements. Astronomers, who make it their business to study the movements of Mars, can see through his evolutions; they know exactly where he is now, and where he will be every night for years and years to come. The people who make the almanacs come to the astronomers and get hints from them as to what Mars intends to do, so that the almanacs announce the positions in which the planet will be found with as much regularity as if he was in the habit of behaving with the orderly propriety of the sun or the moon.

We must not lay all the blame on Mars for the eccentricities of his movements. Our earth is to a very large extent responsible. What we think to be Mars' vagaries are very often to be explained by the fact that we ourselves on the earth are rapidly shifting about and altering our point of view.

I was driving down a pretty country road with a little girl three years old beside me, when I was addressed with the little remark, "Look at the tree going about in the field." Now, you or I, with our longer experience of the world around us, know that it is not the custom of trees to take themselves up and walk about the fields. But this was what this little girl saw, or rather what she thought she saw; and very often what we do see is something very different from what we think we see. We think we see Mars performing all these extraordinary movements, as the little girl thought she saw the tree moving about. But just as that little girl, when she grew to be a big girl, found that what she thought was a tree walking across the field must really have some quite different explanation, so we, too, find that what Mars seems to do is one thing, and what Mars actually does is quite another thing.

Let us see what the little girl noticed. She was looking at the tree, and first she saw it on one side of the house, and then she saw it on the opposite side (Fig. 51). If it had been a cow instead of a tree, of course the natural supposition would have been that the cow had walked. The

Fig. 51.—How the Tree seems to Move about.

little girl may, perhaps, have thought it unusual for a tree to walk, but still she saw the undoubted fact that the tree had shifted to the other side of the house, and therefore, perhaps, remembering what the cow could do, she said the tree had moved.

The little girl entirely omitted to reflect that she herself had entirely changed her position, and hence arose the surprising phenomenon of a tree that could move about. You will understand this, at once, from the two positions of the car here shown. In the first position, as the girl looks at the tree, the dotted line shows the direction of her glance, and the other dotted line shows how the apparent places of the tree and the house have altered. It is her change of place that has accomplished the transformation. Observe also that the tree appeared to her to move in the opposite direction to that in which she is going.

Mars generally appears to move round among the stars from west to east. In fact, if we were viewing him from the sun he would always seem to move in this manner. But at certain seasons our earth is moving very fast past Mars, and this will make him appear to move in the

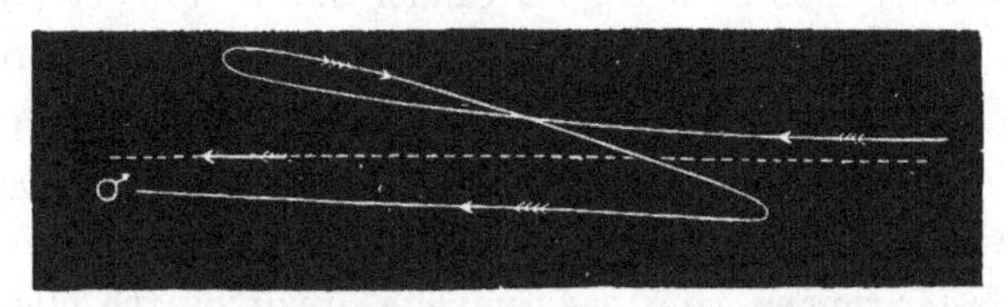

Fig. 52.—A Specimen of the Track of Mars.

opposite direction. This apparent motion is sometimes so much in excess of Mars' real motion, that it may give us an entirely incorrect idea of what he is actually doing.

Thus, notwithstanding that Mars is moving one way, he may appear to us who dwell on the earth to be going in the opposite way. This illusion only happens for a short time, as we pass Mars every two years. The effect on the planet is to make the path he pursues at this time something like that shown in Fig. 52. The planet is nearest to us at the time he is moving in this loop. He is then to be seen at his best in the telescope, so that for every reason it is especially interesting to watch Mars through this critical part of his career.

I want to show you how to make a little calculation that will explain the law by which the seasons when we can see Mars best will follow each other. The period he requires for a voyage round the sun is not quite two years, for that would be 730 days, and Mars only takes 687 days for his journey. It is, however, very nearly true that $1\frac{15}{17}$ years is the period of Mars. Hence, every 32 years Mars will complete 17 rounds. From this we shall be able to see how long it will take after the earth once passes Mars before they pass again. I shall suppose there is a circular course, around which two boys start together to run a race. One of these boys is such a good runner that he will get quite round in 17 minutes; but the other boy can hardly run more than half as quickly, for he will require 32 minutes to complete one circle. Here then is the question. Suppose the two boys to start together: how long will it be before the faster runner gains one complete circuit on the other? By the time the good runner (A) has completed one circuit, the bad runner (B) has only got a little more than half-way. When A has completed his second circuit, he has of course run for twice 17 minutes—that is for 34 minutes. This is two minutes longer than the time B requires to get round once; therefore B is only ahead by a distance which A could cover in about one minute; but B will have advanced during this minute over a distance for which A will require another half-minute, and then in like manner a further quarter, and a half-quarter, and so on. All these added together show that B will not be overtaken until about two minutes after A has completed his second round—that is in 36 minutes altogether.

We can pass from this illustration to the case of the

planet Mars and the earth. The orbit of the earth is traversed in a year, and therefore, after the earth has once passed Mars, which is then, as astronomers would say, in opposition, about two years and the eighth of a year—that is, two years and six or seven weeks—will elapse before Mars is again favourably placed. You will thus see that we need not expect to observe Mars to advantage every year. The distance of the planet from the earth at opposition varies so that some years are more favourable than others.

THE ELLIPSE.

The time has come when I must tell you something about the shapes of the paths in which the earth and the other planets move. Perhaps you will think that I am going to contradict some of the things that I have told you before. I have often represented the orbits of the planets as circles, and now I am going to tell you that this is not correct. The fact is that the paths are nearly circles; but, still, there is some departure from the exact circular shape. Mars, in particular, moves in a path which is more different from a circle than the path of the earth, and consequently it is appropriate to introduce this subject when we are engaged about Mars.

We must first take another lesson in drawing, and the appliances I want you to use for the purpose are very simple. You must have a smooth board and some tacks or drawing-pins, besides paper, pencil, and twine.

We first lay a sheet of paper on the board, and then put in two tacks through the paper and into the board. It does not much matter where we put them in. Next we take a piece of twine and tie the two ends together so as to form a loop, which we pass round the two tacks (Fig. 53). In the loop I place the pencil, and then you see I move it round, taking care to keep the twine stretched. Thus I produce a pretty curve, which we call the ellipse. I must ask all of you to practise this art. Try with different lengths of string, and try using different distances between the tacks. Here are some sketches of two shapes of ellipse

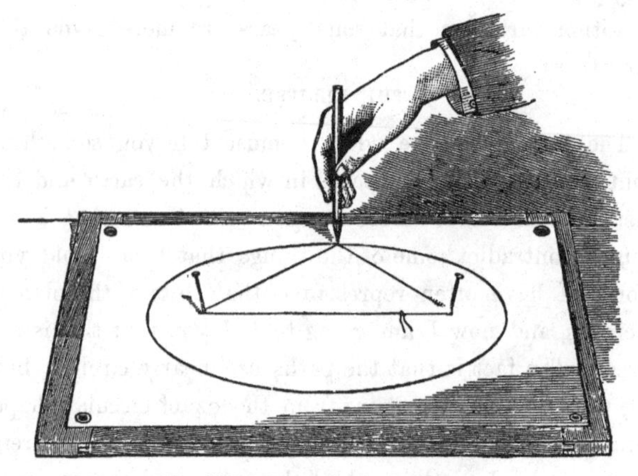

Fig. 53.—How to draw an Ellipse.

and a parabola (Fig. 54). Elliptic curves can be made very nearly circles by putting the two tacks close together, or they can be made very long in comparison with their width. They are all pretty and graceful figures, and are often useful for ornamental work. The ellipse is a pretty shape for beds of flowers in a grass-plot.

The importance of the ellipse to astronomers is greater than that of any other geometrical figure. In fact, all the planets as they perform their long and unceasing journeys round the sun, move in ellipses ; and though it is true that these ellipses are very nearly circles, yet the difference is quite appreciable.

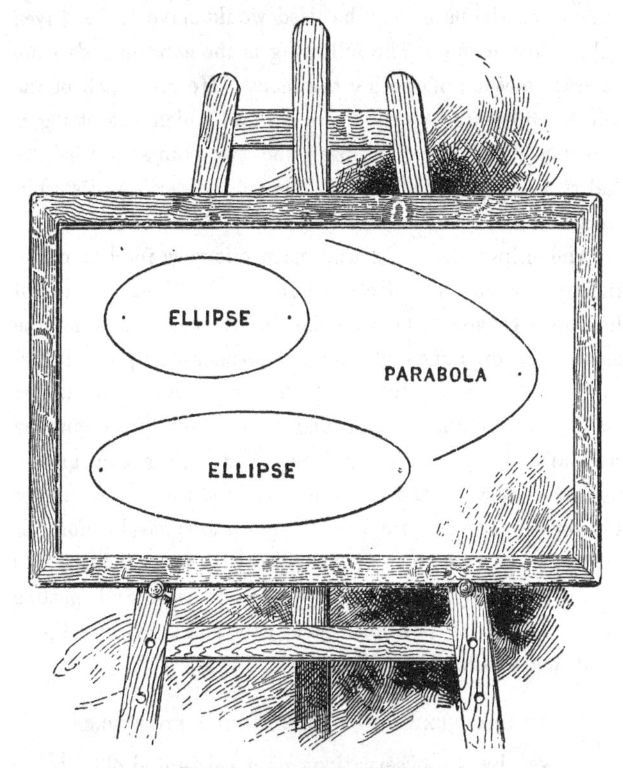

Fig. 54.—Specimens of Ellipses.

It is also important to observe that the sun is not in the centre of the ellipse which the planet describes. The sun is nearer to one end than to the other. And the actual position of the sun must be particularly noted. Suppose that some mighty giant were about to draw an exact path for the earth, or for Mars, of course he would want to have millions of miles of string for producing a big enough curve, and one of the nails that he used would have to be driven right into the sun. The following is the astronomer's more accurate method of stating the facts. He calls each of the points represented by the tacks around which the string is looped a *focus* of the ellipse; the two points together are said to be the *foci*; and as the planet is describing its orbit, the position of the sun will lie exactly at one of the foci.

The ellipse is a curve that nature is very fond of reproducing. From an electric light, a brilliant beam will diverge. If you hold a globe in the beam, and let the shadow fall on a sheet of paper, it forms an ellipse. If you hold the sheet squarely, the shadow is a circle; but as you incline it, you obtain a beautiful oval, or ellipse; and by gradually altering the position, you can get a greatly elongated curve. Indeed, you can produce one of almost any form. The electric light is not indispensable for this purpose; any ordinary bright lamp with a small flame will answer, and by taking different-sized balls and putting them in various positions, you can make many ellipses, great and small.

THE DISCOVERIES MADE BY TYCHO AND KEPLER.

It was by the observations of a celebrated old astronomer, named Tycho Brahe, that the true shape of a planet's path came to be afterwards determined. Tycho lived in days before telescopes were invented. He had few of the excellent contrivances for measuring which we have in our observatories. We shall take a look at this fine old astronomer, as he sits amid his wonderful astronomical machines.

He lived on an island near Copenhagen, and he has given us a picture of himself (Fig. 55), as he is seated with his quaint apparatus, and his assistants around him, busily engaged in observing the heavens. You see the walls of his observatory are decorated with pictures; and one of the great Danish hounds which the King of Denmark had presented to him lies asleep at his feet. I do not think we would now encourage big dogs in the observatory at night. Nor do modern astronomers put on their velvet robes of state, as Tycho was said to have done when he entered into the presence of the stars, as, by so doing, he intended to show his respect for the heavens. Astronomers, nowadays, rather prefer to wear some comfortable coat which shall keep out the cold, no matter what may be its appearance from the picturesque point of view. In this queer old contrivance, you see Tycho Brahe had no actual telescope. He observed through a small opening in the wall, and lest there should be any mistake as to what is going on, you see he is pointing towards it, and giving his three assistants their instructions. The most important work is being done by the man on the right. He is engaged in making the actual observation. But he has no aid from magnifying lenses. All he can do is to slide a pointer up or down till it is just in line with the planet as he sees it through the hole opposite.

On this circle there are a number of marks engraved, and there are numbers placed opposite to the marks; it is by these that the position of the planet is to be ascertained. If the planet is high, then the pointer will be low; and if the planet is low, then the pointer will be high. The

QVADRANS MVRALIS
SIVE TICHONICUS.

Fig. 55.—Tycho in his Observatory.

observer calls out the position when he has found it, and there, you see, is a man ready with a pencil to take down the observation. Notice also the other astronomer who is looking at the clock. He gives the time, which must also be recorded accurately. In fact, the entire process of finding the place of a heavenly body consists in two observations—one from the circle and the other from the clock; so that though Tycho had no telescope to aid his vision, yet the principle on which his work was done was the same as that which we use in our observatories at this moment.

You may think that such a concern would hardly be capable of producing much reliable work. However, Tycho compensated in a great degree for the imperfection of his instrument by the skill with which he used it. He had a noble determination to do his very best. Perseverance will accomplish wonders even with very imperfect means. A great astronomer has said that a skilful observer ought to be able to make valuable measurements with a common cart-wheel!

It was with instruments on the principle of that which I have here shown that Tycho made his celebrated observations of Mars. Week after week, month after month, year after year, did the patient old astronomer track Mars through his capricious wanderings.

Before we try to explain anything, it is of course necessary to ascertain, with all available accuracy, what the thing actually is. Therefore, when we seek to explain the irregular movements of a planet, the first thing to be done is to make a very careful examination of the nature of those irregularities. And this was what Tycho strove to do with the best means at his disposal.

The full benefit of Tycho's work was realised by Kepler when he commenced to search out the kind of figure in which Mars was moving. First he tried various circles, and then he sought, by placing the centre in different positions, to see whether it would not be possible to account thus for the irregularities of the troublesome planet. It would not do; the movement was not circular. This was thought very strange in those days, for the circle was regarded as the only perfect curve, and it was considered quite impossible for a planet to have any motion except it were the most perfect. There was, however, no help for it; so Kepler wisely tried the ellipse, which he considered to be the next most perfect curve to the circle. He continued his long calculations, until at last he succeeded in finding one particular ellipse, placed in one particular position, which would just explain the strange wanderings of our erratic neighbour. It was not alone that the motion of the planet traced out an ellipse; it was further discovered that the sun lies at one of the foci. If the sun were anywhere else, the motion of the planet would have been different from that which Tycho had found it to be.

You must know that this discovery is one of the very greatest that have ever been made in the whole extent of human knowledge. After it had been proved that the orbit of Mars was elliptic, it became plain that the same law must be true for every planet. There are very big planets, and there are small ones; there are planets which move in very large orbits, and there are planets which move in comparatively small ones. In all cases the high road which the planet follows is invariably an ellipse, and the sun is invariably to be found situated at the focus. It is surely interesting to find that these beautiful ellipses which we can draw so simply with a piece of twine and a pencil should be also the very same figures which our great earth and all the other bodies which revolve around the sun are ever compelled to follow.

Kepler also made another great discovery in connection with the same subject. If the planet moved in a circle with the sun in the centre, then there would be very good reason to expect that it would always move at the same speed, for there would be no reason why it should go faster at one place than at another. In fact, the planet would then be revolving always at the same distance from the sun, and every part of its path would be exactly like every other part. But when we consider that the motion is performed in an ellipse, so that the planet is curving round more rapidly at the extremities of its path than in the other parts where the curvature is less perceptible, we have no reason to expect that the speed shall remain the same all round.

We know that the engine driver of a railway train always has to slacken speed when he is going round a sharp curve. If he did not do so, his train would be very likely to run off the line, and a dreadful accident would follow. The engine driver is well aware that the conditions of pace are dependent on the curvature of his line. The planet finds that it, too, must pay attention to the curves; but the extraordinary point is that the planet sometimes acts exactly in the opposite way to the engine driver. The planet puts on its highest pace at one of the most critical curves in the whole journey. There are two specially sharp curves in the planet's path. These are, of course, the two extremities of the ellipse which it follows. The cautious engine driver would, of course, creep round these with equal care, and no doubt the planet goes slowly enough about that end of the ellipse which is farthest from the sun. There his pace is slower than anywhere else; but from that moment onwards the planet steadily applies itself to getting up more and more speed. As it traverses the comparatively straight portion of the celestial road, the pace is ever accelerating until the sharp curve near the sun is being approached; then the velocity gets more and more alarming, until at last, in utter defiance of all rules of engine driving, the planet rushes round one of the worst parts of the orbit at the highest possible speed. And yet no accident happens, though the planet has no nicely laid lines to keep it on the track.

If lines are necessary to save a railway train from destruction, how can we possibly escape when we have no similar assistance to keep us from flying away from the sun and off into infinite space? Kepler has taught us to measure the changes in the speed of the body with precision. He has shown that the planet must, at every point of its long journey, possess exactly the right speed; otherwise everything would go wrong. I daresay you have seen at different points along a line of railway, boards put up here and there, with notices like, "Ten miles an hour." These words are, of course, an intimation to the engine driver that he is not to vary from the speed thus stated. Kepler has given us a law which is equivalent to a large number of caution boards, fixed all round the planet's path, indicating the safe speed for the journey at every stage. It is fortunate for us that the planet is careful to observe these regulations. If the earth were to leave her track, the consequences would be far worse than those of the most frightful railway accident that ever happened. Whichever side we took would be almost equally disastrous. If we went inwards we should plunge into the sun, and if we went outwards we should be frozen by cold.

We owe our safety to the care with which the speed of the earth is prescribed. When near the sun, the earth is pulled inwards with exceptionally strong attraction. We are often told that when a strong temptation seizes us, the wisest thing that we can do is to run away as hard as possible. This is just what the laws of dynamics cause the earth to do at this critical time. She puts on her very best pace, and only slackens when she has got well away from the danger.

The danger that we are exposed to when at the other end of the orbit is of an opposite character. We are then a long way from the sun, and the pull which it can exercise upon the earth is correspondingly lessened. Now, the danger is that we shall escape altogether from the sun's warmth and his guidance. We must therefore give time to the sun to exercise his power, so as to enable the earth to be recalled; accordingly we move as slowly as possible until the sun conquers the earth's disposition to fly off, and we begin to return.

You may remember that when we were speaking about the moon, I showed you how a body might revolve around the earth in a circle under the influence of an attraction towards the earth's centre. So long as the path is really a circle, then the power with which the earth is drawing the body remains the same, because a circle is of the same shape at every part. In a precisely similar way, a body could revolve around the sun in a circle, in which case also the attraction of the sun will remain the same all round. But now we have a very much more difficult case to consider. If the body does not always remain at the same distance, the power of the sun will not be the same at the different places. Whenever the body is near the sun, the attraction will be greater than when the body is farther off. For example, when the distance between the two bodies is doubled, then the pull is reduced to the fourth part of what it was before.

THE DISCOVERIES MADE BY NEWTON.

I have now some great discoveries to talk to you about, which were made by Sir Isaac Newton. He was not an astronomer that looked through a telescope. He used to sit in his study and think, and then he used to draw figures with his pencil, and make long calculations. At last he was able to give answers to the questions: What is the reason why the planet moves in an ellipse? Why should it move in this curve rather than in any other? Why should this ellipse be so placed that the sun lies at one of the foci?

If the planet had run uniformly round its course, Newton would have found his task an impossible one. But I have already explained that the motion was not uniform. I described how the planet hurried along with extra speed at certain points of its path; how it lingered at other points; how, in fact, it never preserved the same rate for even a single minute during the whole journey. Kepler had shown how to make a time-table for the whole journey. In fact, just as a captain on a long voyage keeps a record of each day's run, and shows how to-day he makes

170 miles, and to morrow perhaps 200, and the next day 210, while the day after he may fall back to 120, so Kepler gave rules by which the log of a planet in its voyage round the sun might be so faithfully kept that every day's run would be accurately recorded.

When Newton commenced his work, one of the first questions he had to consider was the following: Suppose that a great globe like a planet, or a small globe like a marble, or an irregular body like an ordinary stone, were to be thrown into space, and was then to be left to follow its course without any force whatever acting upon it, where would it go to?

You may say, at once, that a body under such circumstances will presently fall down to the ground; and so, of course, it will, if it be near the earth. I am not, however, talking of anything near the earth: I want you to imagine a body far off in the depths of space, among the stars. Such a body need not necessarily fall down here, for you see the moon does not fall, and the sun does not fall.

If you were far away from our globe and from all other large globes—so far, indeed, that their attractions were imperceptible—you could try the experiment that I wish now to describe. Throw a stone as hard as ever you can, and what will happen? Of course, when you do it down here, it moves in a pretty curve through the air, and tumbles to the ground; but far off in space, what will the stone do? There will be no such notion as up or down, for though the earth, no doubt, will lie in one particular direction far away, yet there will be other bodies just as large in other directions; and there is no reason why the stone should move towards one of these rather than to another;

in fact, if they are all far enough, as the stars are from us, their attractions will be quite unappreciable. There is, therefore, not the slightest reason why the stone should swerve to one side more than to another. There is no more reason why it should turn to the right than why it should turn to the left. Nor could you throw the stone so as to make it follow a curved path. You can, of course, make it describe a curve while it remains in your hand, but the moment the stone has left your hand, it proceeds on its journey by a law over which you have no control. As the direction cannot be changed towards one side more than towards the other, the stone must simply follow a straight line from the very moment when it is released from your hand.

The speed with which the stone is started will also not change. You might at first think that it would gradually abate, and ultimately cease. No doubt a stone thrown along the road will behave in this way, but that is because the stone rubs against the ground. If you throw a stone across a sheet of ice, then it will run a very long distance before it stops, and all the time it will be moving in a straight line. In this case there is but little loss by rubbing against the ice, because it is so smooth. Thus we see that if the path be exceedingly smooth, the body will run a long way before it stops. Think of the distance that a railway train will still run if, while it is travelling at full speed along a level line, the steam is turned off.

These illustrations all show that if you let a body alone, after having once started it, and do not try to pull it this way or that way, and do not make it rub against things, that body will move on continually in a straight line, and will keep up a uniform speed. We can apply this reasoning to a stone out in space. It would certainly move in a straight line, and would go on and on for ever, without losing any of its pace.

I need hardly tell you that no one has ever been able to try this experiment. In the first place, we reside upon the surface of the earth, and we have no means of ascending into those elevated regions where the stone is supposed to be projected. There is also another difficulty which we cannot entirely avoid, and that arises from the resistance of the air. All movements down here are impeded because the body has to force its way through the air; and in doing so it invariably loses some of its speed. Out in open space there is, of course, no air, and no loss of speed can therefore arise from this cause.

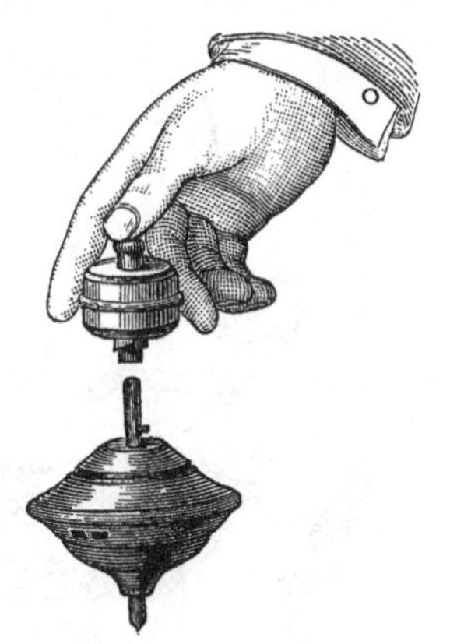

Fig. 56.—The humming-top.

There are, however, several actual experiments by which we can assure ourselves of the general truth. Set a humming-top spinning (Fig. 56); it gradually comes to rest, partly because of the rubbing of its point on the table, and partly because it has to force its way through the air. In fact, the hum of the top that you hear is only produced at the expense of its motion. I have here a much heavier top, and if I set it spinning it will keep up for many minutes, because its weight gives it a better store of power wherewith to overcome the resistance of the air. I remember hearing a story about

Professor Clerk-Maxwell. He had, when at Cambridge, invented one of these large and heavy tops, which would spin for a long time. One evening the top was left spinning on a plate in his room when his friends took their departure, and no doubt it came to rest in due time. Early the next morning, Professor Maxwell, hearing

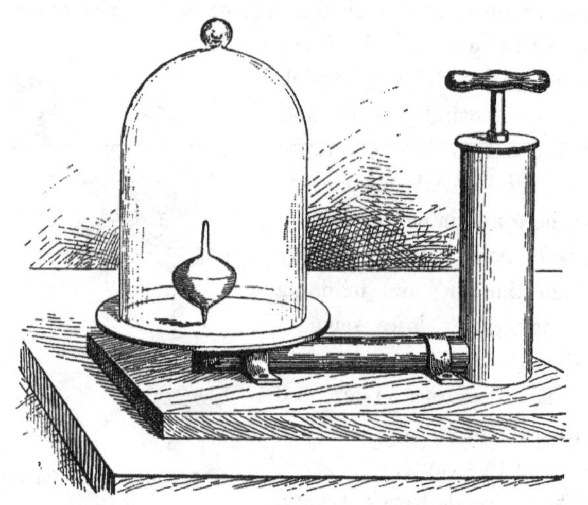

Fig. 57.—To illustrate the first law of motion.

the same friends coming up to his rooms again, jumped out of bed, set the top spinning, and then got back to bed, and pretended to be asleep. He thus astounded his friends, who, of course, imagined that the top must have been spinning all the night long!

If we spin a top under the receiver of an air pump (Fig. 57), it will keep up its motion for a very much longer time after the air has been exhausted than it would under ordinary circumstances. Such experiments prove that the motion of a body will not of itself naturally die out, and that if we could only keep away the interfering forces altogether, the motion would continue indefinitely with unabated speed. What I have been endeavouring to illustrate is called the first law of motion. It is written thus:—

"*Every body continues in its state of rest or of uniform motion in a straight line, except in so far as it may be compelled by impressed forces to change that state.*"

I would recommend you to learn this by heart. I can assure you it is quite as well worth knowing as those rules in the Latin Grammar with which many of you, I have no doubt, are acquainted. The best proof of the first law of motion is derived, not from any experiments, but from astronomy. We make many calculations about the movements of the sun, the moon, the stars, and then we venture on predictions, and we find those predictions verified. Thus we had a transit of Venus across the sun in 1882, and every astronomer knew that this was going to occur, and many went to the ends of the earth so as to see it favourably. Their anticipations were realised; they always are. Astronomers make no mistakes in these matters. They know that there will be another transit of Venus in the year 2004, but not sooner. The calculations by which these accurate prophecies are made involve this first law of motion; and as we find that such prophecies always come true, we know that the first law of motion must be true also.

Newton knew that if a planet were merely left alone in space, it would continue to move on for ever in a straight line. But Kepler had shown that the planet did not move in a straight line, but that it described an ellipse. One conclusion was obvious. There must be some force acting upon the planet which pulls it away from the straight line it would otherwise pursue. We may, for the sake of illustration, imagine this force to be applied by a rope attached to the planet, so that at every moment it is dragged by some unseen hand. To find the direction this rope must have, we take the law of Kepler, which explains the rules according to which the planet varies its speed. I cannot enter into the question fully, as it would be too difficult for us to discuss now. I should have to talk a great deal more about mathematics than would be convenient just at present; but I think you can all understand the result to which Newton was led. He showed that the rope must always be directed towards the sun. In other words, suppose that there was no sun, but that in the place which it occupied there was a strong enough giant constantly pulling away at the planet, then we should find that the speed of the planet would alter just in the way it actually does. Thus we learn that some force must reside in the sun by which the planet is drawn, and this force is exerted, although there is no visible bond between the sun and the planet.

There is another fact to be learned about the sun's attraction, and this time we obtain it by knowing the shape of the curve followed by the planet. The laws by which the planet's speed is regulated prove that the force emanates from the sun. We shall now learn much more when we take into account that the path of the planet is an ellipse, of which the sun lies at the focus. We have said nothing as yet as to the magnitude of the pull which is being exerted by the sun. Is that pull to be always the same, or is it to be greater at some times than at other times? Newton showed that no ellipse other than a circle could be described, if the pull from the sun were always the same. Its magnitude must be continually changed, and the nearer the planet lies to the sun, the more vehement is the pull it receives. Newton laid down the exact law by which the force on the planet at any one position of its path could be compared with the force at any other position. Let us suppose that the planet is in a certain position, and that it then passes into a second position, which is twice as far from the sun. The pull upon the planet at the shorter distance is not only greater than the pull at the longer distance, but it is actually four times as much. Stating this result a little more generally, we assert, in the language of astronomers, that the *attraction varies inversely as the square of the distance.* If this law were departed from, then I do not say that it would be impossible for the planet to revolve around the sun in some fashion, but the motion would not be performed in an ellipse described around the sun in the focus.

You see how very instructive are the laws which Kepler discovered. From the first of them we were able to infer that the sun attracts the planets; from the second, we have learned how the magnitude of the attracting force varies.

The true importance of these great discoveries will be manifest when we compare them with what we have already learned with regard to the movements of the moon. As the moon revolves around the earth it is held by the earth's attraction, and the moon follows a path which, though nearly a circle, is really an ellipse. This orbit is described around the earth just as the earth describes its path around the sun. That law by which a stone falls to the ground in consequence of the earth's attraction is merely an illustration of a great general principle. Every body in the whole universe attracts every other body.

Think of two weights lying on the table. They no doubt attract each other, but the force is an extremely small one—so small, indeed, that you could not measure it by any ordinary appliance. One or both of the attracting masses must be enormously big if their mutual gravitation is to be readily appreciable. The attraction of the earth on a stone is a considerable force, because the earth is so large, even though the stone may be small. Imagine a pair of colossal solid iron cannon balls, each 53 yards in diameter, and weighing about 417,000 tons. Suppose these two globes were placed a mile apart, the pull of one of them on the other by gravitation would be just a pound weight. Notwithstanding the size of these masses, the hand of a child could prevent any motion of one ball by the attraction of the other. If, however, they were quite free to move, and there was absolutely no friction, the balls would begin to draw together; at first they would creep so slowly that the motion would hardly be noticed. The pace would no doubt continue to improve slowly, but still not less than three or four days must elapse before they will have come together.

By the kindness of Professor Dewar, I am enabled to exhibit a contrivance with which we can illustrate the motion of a planet around the sun. Here is a long wire suspended from the roof of this theatre, and attached to its lower end is an iron ball, made hollow for the sake of lightness. When I draw the ball aside, it swings to and fro with the regularity of a great pendulum. But when I place a powerful magnet in its neighbourhood (Fig. 58), you see that as soon as the ball gets near the magnet it is violently drawn to one side, and follows a curved path. This magnet may be taken to represent the sun, while the ball is like our earth, or any other planet, which would

Fig. 58.—The Effect of Attraction.

move in a straight line were it not for the attraction of the sun which draws the body aside.

THE GEOGRAPHY OF MARS.

It remains to say something with respect to the geography of our fellow-planet, a subject which seems all the more interesting because Mars is so like the earth in many respects. We require a pretty good telescope for the purpose of seeing him well, but when a sufficient instrument

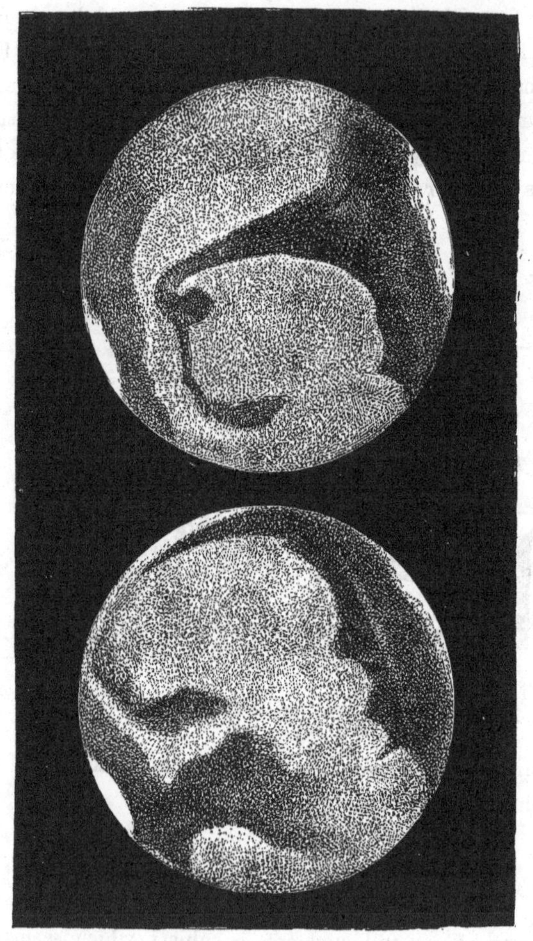

Fig. 59.—Views of Mars.

is directed to the planet, a beautiful picture of another world is unfolded (Fig. 59). There are many maikings visible on his surface, but we must always remember that even with our most powerful telescopes the planet still appears a long way off.

Under the most favourable circumstances, Mars is at least one hundred times as far from us as the moon. But we know that an object on the moon must be as large as St. Paul's Cathedral if it is to be visible in our telescopes. An object on Mars must be, therefore, at least one hundred times as broad and one hundred times as long as St. Paul's Cathedral if it is to be discernible by astronomers here. We can, therefore, only expect to see the broad features of our fellow-planet. Were we looking at our earth from a similar distance, and with equally good telescopes, the continents and oceans, and the larger seas and islands, would all be large enough to be conspicuous. It is, however, doubtful whether they could be ever properly revealed through the serious impediment to vision which our atmosphere would offer.

It fortunately happens that the surface of Mars is only obscured by clouds to a very trifling extent, and we are thus able to see a panorama of our neighbouring globe laid before us. Mars is not nearly as large as our earth, the diameters of the two bodies being nearly as two to one. It follows that the number of acres on the planet are only a quarter of the number of acres on the earth. Careful telescopic scrutiny shows that the chief features which we see on Mars are of a permanent character. In this respect Mars is much more like the moon than the sun. The latter presents to us merely glowing vapours, with hardly more permanence than is possessed by the clouds in our own sky. On the other hand, the entire absence of

M 2

clouds from the moon makes all its features quite constant. Most of the visible features on Mars are also invariable; though occasionally it would seem that the climate produces some changes.

We first notice that there are differently coloured parts on Mars. The darkish or bluish regions are usually spoken of as being seas or oceans; though we should be going beyond our strict knowledge were we to assert that water is actually found there. Look at the horn-shaped object in the centre of the upper picture in Fig. 59. We call it the Kaiser Sea, and it is so strongly marked that even in a small telescope it can be often seen. You must not, however, always expect to notice this feature when you look at Mars with a telescope, for the planet turns round and round. We can make a globe representing Mars. On this are to be depicted the Kaiser Sea and the remaining characteristic objects. But as we turn the globe around, the opposite side of the planet is brought into view, and other features are exposed like those represented in the lower figure. Mars requires 24 hours, 37 minutes, 22·7 seconds to complete a single rotation. It is somewhat remarkable that this only differs from the earth's period of rotation by a little more than half an hour.

Mars contains what we call continents as well as oceans, and we also find there lakes and seas and straits. These objects are indicated in the drawings that are here represented. But the most striking feature which the planet displays are the marvellous white regions, which are seen both at its North Pole and at its South Pole (Fig. 60). If we were able to soar aloft above our earth, and take a bird's-eye view of our own polar regions, we should see a white cap at the middle of the arctic circle. This appearance would be produced by the eternal ice and snow. It would increase during the long, dark winter, and then be somewhat reduced by melting during the continuously bright summer. Though we cannot thus see our earth, yet we can sometimes observe one Pole of Mars and sometimes the other, and we find each of these Poles crowned with a dense white cap, which increases during the severity

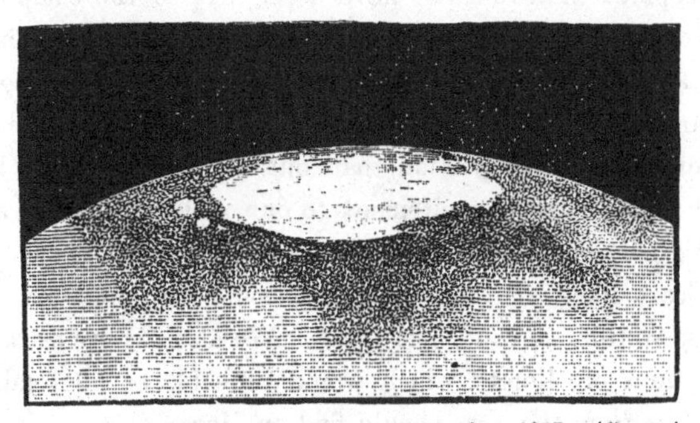

Fig. 60.—The South Pole of Mars, September, 1877. (Green.)

of Mars' winter, and which declines again in some degree with the warmth of the ensuing summer.

Sketches of Mars have been made by many astronomers; among them we may mention Mr. Green, who made a beautiful series of pictures at Madeira in 1877. These may be supplemented by the drawings of Mr. Knobel in 1884, when the opposite Pole of Mars was turned to view. The drawings show the Polar snows of Mars, and there seem to be some elevated districts in the arctic regions which retain a little patch of snow after the main body of the ice cap has shrunk within its summer limits. An interesting case of this kind is shown in Fig. 60, which has been copied from one of Mr. Green's drawings.

It has lately been surmised that the continents on Mars are occasionally inundated by floods of water. There are also indications of clouds hanging over the Martian lands, but the inhabitants of that planet, in this respect, escape much better than we do. A certain amount of atmosphere always surrounds Mars, though it is much less copious than that we have here. As to the composition of this atmosphere we know nothing. For anything we can tell, it might be a gas so poisonous that a single inspiration would be fatal to us; or if it contained oxygen in much larger proportion than our air does, it might be fatal from the mere excitement to our circulation which an over-supply of stimulant would produce. I do not think it the least likely that our existence could be supported on Mars, even if we could get there. We also require certain conditions of climate, which would probably be all totally different from those we should find on Mars.

THE SATELLITES OF MARS.

When Mars appeared in his full splendour in 1877, he was for the first time honoured with the notice of instruments capable of doing him justice; I do not, however, mean that in former apparitions he was not also carefully observed, but a great improvement had recently taken place in telescopes, and it was thus under specially favourable auspices that his return was welcomed in 1877. This year will be always celebrated in astronomical history for a beautiful discovery, made by Professor Asaph Hall, the illustrious astronomer at Washington.

Before I can explain what this discovery was, I must have a little talk about moons, or satellites as we often call them. You know that we have one moon, which is constantly revolving round the earth, and accompanies the earth in its long voyage round the sun. But the earth is only a planet, and there are many other planets which are worlds like ours. It is natural to compare these worlds together, and as we have one moon, why should not the other planets also have moons? If there are children in one house in a square, why should there not be children in the other houses? We find that some of the other planets have satellites, but they do not seem to be distributed very regularly. In fact, they are almost as capriciously allotted as the children would be in eight houses that you might take at random.

In Number One there lives an old bachelor, and in Number Two a single lady. These are Mercury and Venus, and of course there are no children in either of these houses. Number Three is inhabited by old mother Earth, and she has got a fine big son, called the Moon. Number Four is a nice little house inhabited by Mars. There are to be found a pair of little twins, and nimble creatures they are too. Number Five is a great mansion. A very big man lives here, called Jupiter, with four robust sons and daughters that everybody knows. I fancy they must go to many dancing parties, for every night they may be seen whirling round and round. Number Six is also a fine big house, though not quite so big as Number Five, but larger than any of the others. It is inhabited by Saturn, and contains the biggest family of all. There are eight sons and daughters here, but they are not nearly so sturdy as Jupiter's children; in fact, the young

Saturns do not make much display, and some of them are so delicate that they hardly ever appear. Number Seven is also a fine large house; but Uranus, who lives there, is such a recluse that unless you carefully keep your eye on his house, you will hardly ever catch a glimpse of him. There are four children in that house, I believe, but we hardly know them. They move in circles of their own, and apparently have seen a good deal of trouble. Only one more house is to be mentioned, and that is Number Eight, inhabited by Neptune. It contains one child, but we are hardly on visiting terms with this household, and we know next to nothing about it.

Before 1877, Mars appeared to be in the same condition as Venus or Mercury—that is, devoid of the dignity of attendants. There was, however, good reason for thinking that there might be some satellites to Mars, only that we had not seen them. You see that, as Number Three had one child, and Numbers Five, Six, Seven, and Eight had each one, or more than one, it seemed hard that poor Number Four should have none at all. It was, however, certain that if there were any satellites to Mars, they must be very tiny things; for if Mars had even one considerable moon, it must have been discovered long ago.

On the memorable occasion in 1877, Professor Hall discovered that the ruddy planet Mars was attended, not alone by one moon, but by two. Their behaviour was most extraordinary. It appeared to him at first almost as if one of these little moons was playing at hide-and-seek. Sometimes it would peep out at one side of the planet, and sometimes at the other side. I have here a picture (Fig. 61) which shows how these moons of Mars revolve.

That is the globe of the planet himself in the middle, and he is turning round steadily in a period which is nearly the same as our day. But the remarkable point is that the inner of the moons of Mars runs round the planet

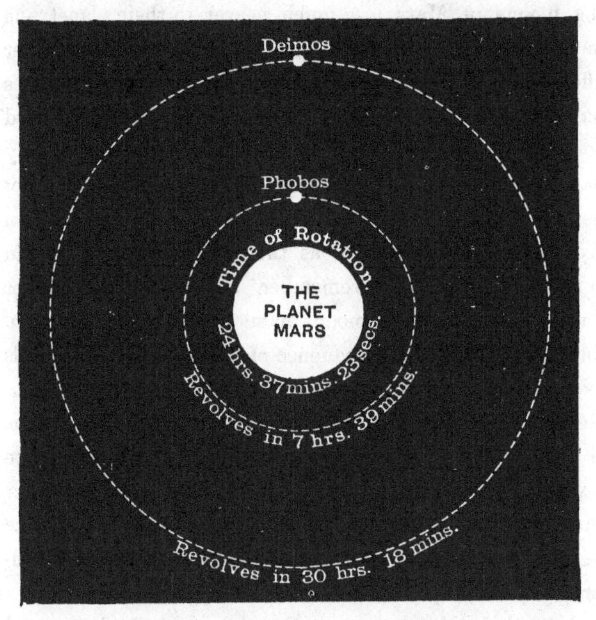

Fig. 61.—Mars and his two Satellites.

in 7 hours 39 minutes. It would seem very strange in our sky if we had a little moon which rose in the west instead of in the east, and which galloped right across the heavens three times every day. The outer moon takes a more leisurely journey, for he requires 30 hours 18 minutes to complete a circuit. If for no other reason than to see these wonderful moons, it would be very interesting to visit Mars.

The satellites of this planet are in contrast to our moon. In the first place, our moon takes 27 days to go round the earth, and is comparatively a long way off. The moons of Mars are much nearer to their planet, and they go round much more quickly. There is also another difference. The moons of Mars are much smaller bodies than our moon. If we represent Mars by a good-sized football, his moons, on the same scale, would be hardly so big as the smallest-sized grains of shot. Does it not speak well for the power of telescopes in these modern days that objects so small as the satellites of Mars should be seen at all? You remember, of course, that neither Mars himself nor his moons have any light of their own. They shine solely in consequence of the sunlight which falls upon them. They are merely lighted like the earth itself, or like the moon. The difficulty about observing the satellites is all the greater because they are seen in the telescope close to such a brilliant body as Mars. The glare from the bright planet is such that when we want to see faint objects like the satellites we have to hide the planet, so as to get a comparatively dark space in which to search.

Now that they know exactly what to look for, a good many astronomers have observed the satellites of Mars. A superb telescope is nevertheless required. And, in fact, you could not find a better test for the excellence of an instrument than to try if it will show these delicate objects. But do not imagine that merely having a good telescope and a clear sky is all that is requisite for making astronomical discoveries. You might just as well say that by putting a first-rate cricket-bat into any man's hands you will ensure his making a grand score. Every boy knows that the bat does not make the cricketer, and I can assure him that neither will the telescope make the astronomer. In both cases, no doubt, there is some element of luck. But of this you may be certain: that as it is the man that makes the score, and not the bat, so it is the astronomer that makes the discovery, and not his telescope.

Deimos and Phobos were the names of the two personages, according to Homer, whose duty it was to attend on the god Mars, and to yoke his steeds. A conclave of classical scholars and astronomers appropriately decided that Deimos and Phobos must be the names of the two satellites to the planet which bears the name of Mars.

HOW THE TELESCOPE AIDS IN VIEWING FAINT OBJECTS.

We have been hitherto talking about large planets, which, if not as big as our earth, are at least as big as our moon. But now we have to say a few words about a number of little planets, many of them being so very small, that a million rolled together would not form a globe so big as this earth. These little objects you cannot see with your unaided eye, and even with a telescope they only look like very small stars.

I have often been asked why it is that a telescope enables us to see objects, both faint and small, which our unaided eyes fail to show. Perhaps this will be a good opportunity to say a few words on the subject. I think we can explain the utility of the telescope by examining our own eyes. The eye undergoes a remarkable transformation when its owner passes from a dark room into a brilliantly lighted room (Fig. 62). Here you see two views of an eye, and you notice the great difference between them. They are not intended to be the eyes of two distinct people; they are merely two conditions of the same eye. They are intended to illustrate two different states of the eye of a collier. The right shows his eye when he is above ground in bright daylight; the left is his eye when he has gone down the coal-pit to his useful work in the dark regions below. I remember when I went down a coal-pit I was lowered down a long shaft, and when I reached the bottom I was handed a safety-lamp.

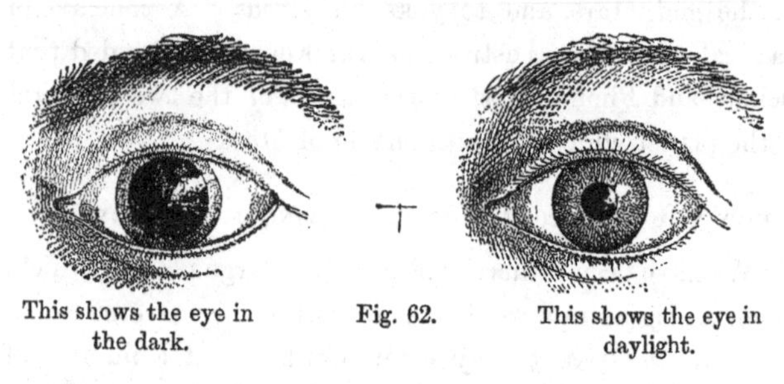

This shows the eye in the dark. Fig. 62. This shows the eye in daylight.

The gloom was such, that at first I found some little difficulty in guiding my steps, but the capable guide beside me said in an encouraging voice, "You will be all right, sir, in a few moments, for you will get your pit-eyes." I did get my "pit-eyes," as he promised, and was able to see my way along sufficiently to enjoy the wonderful sights that are met with in the depths below.

The change that really came over my eyes is that which these two pictures illustrate: the black, round spot in the centre is an opening covered with a transparent window, by which light enters the eye; the black spot is called the pupil, and nature has provided a beautiful contrivance by which the pupil can get larger or smaller, so as to make vision agreeable. When there is a great deal of light we limit the amount that enters by contracting the pupil, so as to make the opening smaller. Thus the picture with the small pupil represents the state of the collier's eye when he is above ground in bright sunlight. When he descends the pit, where the light is very scanty, then he wants to grasp as much of it as ever he can, and consequently his pupil enlarges so as to make a wider opening, and this is what he calls getting his "pit-eyes."

But you need not go down a coal-mine to see the use of the iris—for so that pretty membrane is called which surrounds the pupil. Every time you pass from light into darkness the same thing can be perceived. When we turn down the lights in the room, so that we are in comparative darkness, our pupils gradually expand. As soon as the lights are turned up again, then our pupils begin to contract. Other animals have the same contrivance in their eyes. You may notice in the Zoological Gardens how quickly the pupil of the lion contracts when he raises his eyes to the light. The power of rapidly changing the pupil might be of service to a beast of prey. Imagine him crouching in a dense shade to wait for his dinner; then of course the pupil will be large from deficiency of light: but when he springs out suddenly on his victim, in bright light, it would surely be of advantage to him to be able at once to see clearly. Accordingly his pupil adjusts itself to the altered conditions with a rapidity that might not be necessary for creatures of less predaceous habits.

These changes of the pupil explain how the telescope aids our eyes when we want to discern very faint objects, like the little planets. Such bodies are not visible to the unaided eyes, because our pupils are not large enough to grasp sufficient light for the purpose. Even when they are opened to the utmost, we want something that shall enable them to open wider still. We must therefore resort to some device which shall have an effect equivalent to an enlargement of the pupil far beyond the limits that nature has actually assigned to it. What we want is something like a funnel which shall transform a large beam of rays into a small one. I may explain what I mean by the following illustration: Suppose that it is raining heavily, and that you want to fill a bucket with water. If you merely put the bucket out in the middle of a field, it will never be filled; but bring it to where the rain-shoot from a house-top is running down, and then your bucket will be running over in a few moments. The reason of course is that the broad top of the house has caught a vast number of drops and brought them together into the narrow shoot, and so the bucket is filled. In the same way the telescope gathers the rays of light that fall on the object-glass, and condenses them into a small beam which can enter the eye. We thus have what is nearly equivalent to an eye with a pupil as big as the object-glass. Thus the effect of a grand telescope amounts to a practical increase of the pupil from the size of a threepenny-piece up to that of a dinner-plate, or even much larger still.

THE ASTEROIDS OR SMALL PLANETS.

An asteroid is like a tiny star, and in fact the two bodies are very often mistaken. If we could get near to the objects, we should see difference enough. We should find the asteroid to be a dark planet like our earth, lighted only by the rays from the sun. The star, small and faint though it may seem, is itself a bright sun, at such a vast distance that it is only visible as a small point. The star is millions of times as far from us as the planet is, and utterly different in every respect.

It is a curious fact that the planets should happen to resemble the stars so closely. We can find an analogous fact in quite another part of nature. In visiting a good entomological collection, you will be shown some of the wonderful leaf-like insects. These creatures have wings, exactly formed to imitate leaves of trees, with the stalks and the veins completely represented. When one of these insects lies at rest, with its wings folded, among a number of leaves, it would be almost impossible to penetrate the disguise. This mimicry is no doubt an ingenious artifice by which the birds or other enemies that want to eat the insects shall be deceived. There is, however, one test which the cunning bird could apply: the leaves do not move about of their own accord, but the leaf insects do. If therefore the bird will only have the patience to wait, he will see a pair of the seeming leaves move, and then the deception will be to him a deception no longer, and he will gobble up the poor insect.

In our attempts to discover the planets we experience just the same difficulties as the insect eating bird. Wide as is the true difference between a planet and a star, there is yet such a seeming resemblance between them that we are often puzzled to know which is which. The planets imitate the stars so successfully, that when one of them is presented to us among myriads of stars it is impossible for us to detect the planet by its appearance. But we can be cunning—we can steadily watch, and the moment we find one of these star-like points beginning to creep about we can pounce upon it. We know by its movements that it is only disguised as a star, but that it is really one of the planets.

It is not always easy to discover the asteroids even by this principle, for unfortunately these bodies move very slowly. If you have a planet in the field of view, it will steal along so gradually, that an hour or more must have elapsed before it has shifted its position with respect to the neighbouring stars to any appreciable extent. The search for such little planets is therefore a tedious one, but there are two methods of conducting it: the new one, which is only just coming into use, and the old one. I shall speak of the old one first.

Although the body's motion is so slow, yet when sufficient time is allowed, the planet will not only move away from the stars close by, but will even journey round the entire heavens. The surest way of making the discovery is to study a small part of the heavens now and to examine the same locality again months or years afterwards. Memory will not suffice for this purpose. No one could recollect all the stars he saw with sufficient distinctness to be confident as to whether the field of the telescope on the second occasion contained either more or fewer stars than it did on the first. The only way of doing this work is to draw a map of all the stars very carefully. This is a tedious business, for the stars are so numerous, that even in a small part of the heavens there will be many thousands of stars visible in the telescope. All of these will have to be entered faithfully in their true places on the map. When this has been done the map must be laid aside for a season, and then it is brought out again and compared with the sky. No doubt the great majority of the objects will be found just as they were before. These are the stars, the distant suns, and our concern is not at present with them. Sometimes it will happen that an object marked on the first map has left a vacant place on the second. This, however, does not help us much, for, whatever the object was, it has vanished into obscurity, and a new planet could hardly be discovered in this way. But sometimes it will happen that there is a small point of light seen in the second map which had no corresponding point in the first map. Then, indeed, the expectation of the astronomer is aroused; he may be on the brink of a discovery. Of course, he watches accurately the little stranger. It might be some star that had been accidentally overlooked when forming the map, or it might possibly be a star that has become bright in the interval. But there is a ready test: is the body moving? He looks at it very carefully, and notes its position with respect to the adjacent stars. In an hour or two his suspicions may be confirmed; if the object be in motion, then it is really a planet. A few further observations, made on subsequent days, will show the path of the planet. But the astronomer must assure himself that the object is not one of the planets that have been already found before he announces his discovery.

The new method of searching for small planets, which is only just coming into use, is a very beautiful one; and it seems to hold out the prospect of making such discoveries more easily in the future than has been possible in the past.

We can take photographs of the heavenly bodies by

adjusting a sensitive plate in the telescope so that the images of the objects we desire shall fall upon it. The method will apply to very small stars, if by excellent clock-work and careful guiding we can keep the telescope constantly pointing to the same spot until the stars have had time to imprint their little images. Thus we obtain a map of the heavens, made in a thoroughly accurate manner. Indeed, the delicacy of photography for this purpose is so great, that the plates show many stars which cannot be seen with even the greatest of telescopes. Suppose that a little planet happened to lie among the stars which are being photographed. All the time that the plate is being exposed the wanderer is, of course, creeping along, and after an hour (exposures even longer are often used), it may have moved through a distance sufficient to ensure its detection. The plate will, therefore, show the stars as points, but the planet will betray its presence by producing a streak.

The asteroids now known number nearly 300. Out of this host a few afford some information to the astronomer, but the majority of them are objects possessing individually only the slightest interest. No small planet is worth looking at as a telescopic picture. We should consider that asteroid to be a large one which possessed a surface altogether as great as England or France. Many of these planets have a superficial extent not so large as some of our great counties. A globe which was just big enough to be covered by Yorkshire—if you could imagine that large county neatly folded round it—would make a very respectable minor planet.

We know hardly anything of the nature of these small worlds, but it is certain that any living beings they could support must have a totally different nature from the creatures that we know on this earth. We can easily prove this by making a calculation. I shall suppose a small planet one hundred miles wide, its diameter having, therefore, the eightieth part of the diameter of the earth. If we were landed on such a globe, we should be far more puzzled by the extraordinary lightness of everything than we should be in the similar case to which I referred of the moon (p. 120). If we supposed the planet to be constituted of materials which had the same density as those of which the earth is made, then every weight would be reduced to the eightieth part of what it is here.

There would be one curious consequence of residence on such a globe. We have heard of attempts to make flying machines, or to provide a man with wings by which he shall soar aloft like the birds. All such contrivances have hitherto failed. It may be possible to make a pair of wings by which a man can fly down, but it is quite another matter when he tries to fly up again. Suppose, however, we were living on a small planet, it would be perfectly easy to fly there, for as our bodies would only seem to weigh a couple of pounds, we ought to be able to flap a pair of wings strong enough to overcome so trivial a force. I should, however, add that this is on the supposition that the atmosphere has the same density as our own.

Life on these small planets would indeed be extraordinary. Let us take, for example, Flora, and see how a game of lawn tennis on that body would be managed. The very lightest blow of the racket would drive the ball a prodigious distance before it could touch the ground; indeed, unless the courts were about half a mile long, it would be impossible to serve any ball that was not a fault.

Nor is there any great exertion necessary for playing lawn tennis on Flora, even though the courts are several hundred acres in extent. As a young lady ran to meet the ball and return it, each of her steps might cover a hundred yards or so with perfect grace; and should she have the misfortune to get a fall, her descent to the ground would be as gentle as if she was seeking repose on a bed of the softest swandown.

These little planets cluster together in a certain part of our system. Inside are the four inner planets, of which we have already spoken, outside are the four outer planets, of which we have soon to speak. Between these two groups there was a vacant space. It seemed unreasonable that where there was room for planets, planets should not be found. Accordingly the search was made, and these objects were discovered. Even at the present day, more and more are being constantly added to the list.

We occasionally get information from these little bodies; for in their revolutions through the solar system, they sometimes pick up scraps of useful knowledge, which we can elicit from them by careful examination. For example, one of the most important problems in the whole of astronomy is to discover the sun's distance. I have already mentioned one of the ways of doing this, which is given by the transit of Venus. We never like to rely on a single method; we are therefore glad to discover any other means of solving the same problem. This it is which the little planets will sometimes do for us. Juno on one occasion approached very close to the earth, and astronomers in various parts of the globe observed her at the same time. When they compared their observations together they measured the sun's distance. But I am not going to trouble you now with a matter so difficult as this. Suffice it to say, that for this, as for all similar investigations, they were constrained to use the very same principle as that which we illustrated in Fig. 5.

Let me rather close this lecture by reminding you that we have here been considering only the lesser members of the great family which circulate round the sun, and that we shall speak in our next lecture of the giant members of our system.

LECTURE IV.

THE GIANT PLANETS.

Jupiter, Saturn, Uranus, Neptune—Jupiter—The Satellites of Jupiter—Saturn—The Nature of the Rings—William Herschel—The Discovery of Uranus—The Satellites of Uranus—The Discovery of Neptune.

JUPITER, SATURN, URANUS, NEPTUNE.

OUR lecture to-day ought to make us take a very humble view of the size of our earth. Mercury, Venus, and Mars may be regarded as our peers, though we are slightly larger than Venus, and a good deal larger than Mercury or Mars; but all these four globes are utterly insignificant in comparison with the gigantic planets which lie in the outer parts of our system. These great bodies do not enjoy the benefits of the sun to the same extent that we are permitted to do; they are so far off that the sun's rays become greatly enfeebled before they can traverse the distance; but the gloom of their situation seems to matter but little, for it is highly improbable that these bodies could be inhabited. Though they get but little of the sun's heat, they seem to have ample heat of their own.

A view of parts of the paths of these four great planets is shown in Fig. 63. The innermost is Jupiter, which completes a circuit in about twelve years; then comes Saturn, revolving in an orbit so great that twenty-nine years and a half is required before the complete journey is finished. Still further outside is Uranus, which has a longer journey than Saturn, moves so much more slowly, that a man would have to live to the ripe old age of eighty-four

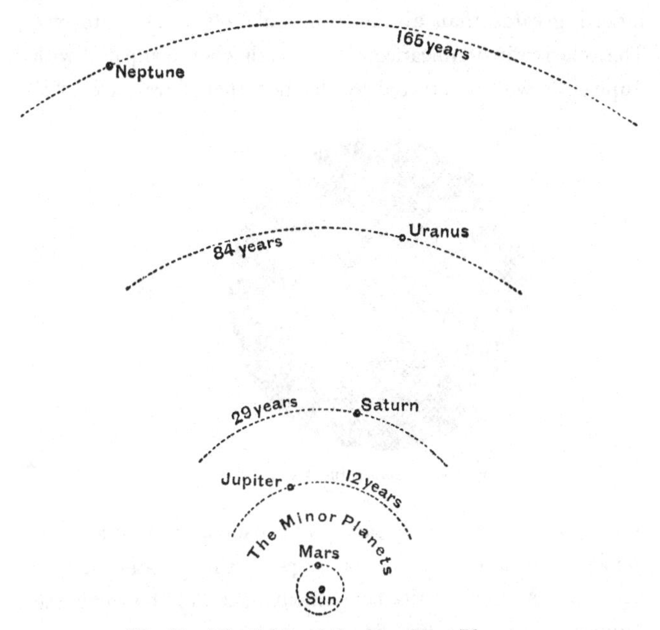

Fig. 63.—The Orbits of the four Giant Planets.

if a complete revolution of Uranus was to be accomplished during his lifetime. At the boundary of our system revolves the planet Neptune, and though it is a mighty globe, yet we cannot see it without a telescope. It is otherwise invisible for two reasons: first of all, because it is so far from the sun that the light which illuminates it is excessively feeble, and, secondly, because it is so far from us that its brilliancy, such as it is, is largely reduced.

JUPITER.

Of all these bodies Jupiter is by far the greatest; he is, indeed, greater than all the other planets rolled into one. The relative insignificance of the earth when compared with Jupiter is well illustrated by the fact that if we took 1,200

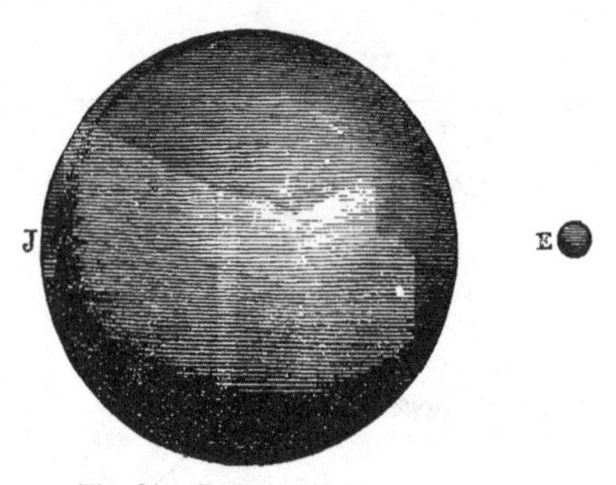

Fig. 64.—Jupiter and Earth compared.

globes each as big as our earth, and made them into a single globe, it would only be as large as the greatest of the planets. A view of the comparative sizes of the earth and Jupiter is shown in Fig. 64.

Here is a picture of Jupiter (Fig. 65), as seen through the telescope. First, you will notice that the outline of the planet's shape is not circular, for it is plain that the vertical diameter in this picture is shorter than the horizontal one; in fact, Jupiter is flattened at the poles and bulges out at the equator, so that a section through the poles is an ellipse. Jupiter is turning round rapidly on his axis, and this will account for the protuberance. We find that the planet has assumed almost the same form as if it were actually a liquid. This we can illustrate by a globe of oil which is poised in a mixture of spirits of wine and water so carefully adjusted that the oil has no tendency

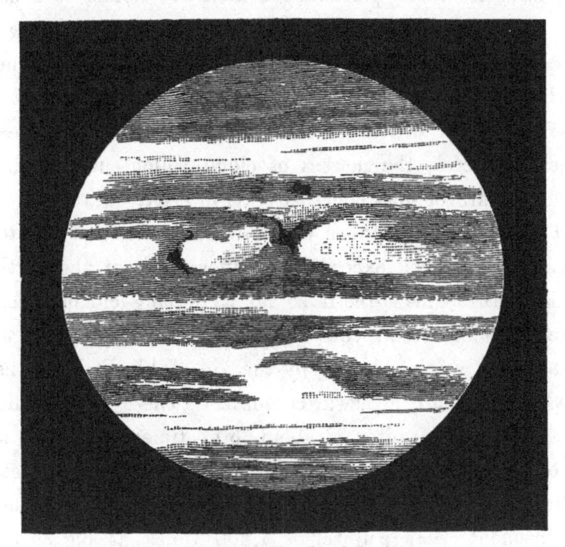

Fig. 65.—The Clouds of Jupiter.

to rise or fall. As we make the globe of oil rotate, which we can easily do by passing a spindle through it, we see that it bulges out into the form that Jupiter as well as other planets have taken.

On the picture of the planet you will see shaded bands. These are constantly changing their aspect, and for a double reason. In the first place, they change because Jupiter is turning around so quickly that in five hours the whole side of the planet which is towards us has been turned away. In another five hours the original side of the globe will be back again, for the entire rotation occupies about ten hours, or 9h. 55m. 21s. precisely.

But these bands are themselves not permanent objects. They have no more permanence than the clouds over our own sky. Sometimes Jupiter's clouds are more strongly marked than on other occasions. Sometimes, indeed, they can hardly be seen at all. It is from this we learn that those markings which we see when we look at the great planet are merely the masses of cloud which surround and obscure whatever may constitute his interior.

There is a circumstance which demonstrates that Jupiter must be an object exceedingly different from the earth, though both bodies agree in so far as having clouds is concerned. What would you think when I tell you that we are able to weigh Jupiter by the aid of his little moons, of which I shall afterwards speak? These little bodies inform us that Jupiter is about 300 times as heavy as our earth, and we have no doubt about this, for it has been confirmed in other ways. But we have found by actual measurement that Jupiter is 1,200 times as big as the earth, and, therefore, if he were constituted like the earth, he ought to be 1,200 times as heavy. This is, I think, quite plain; for if two cakes were made of the same material, and one contained twice the bulk of the other, then it would certainly be twice as heavy. If of two balls of iron one have twice the bulk of the other, then, of course, it is twice as heavy. But if a ball of lead have twice the bulk of a ball of iron, then the leaden ball would be more than twice as heavy as the iron, because lead is a heavier material than iron. In the same way, the weights of the earth and Jupiter are not what we might expect from their relative sizes. If the two bodies were made of the same materials and in the same state, then Jupiter would be certainly four times as heavy as we find him to be. We are, therefore, led to the belief that Jupiter is not a solid body, at least in its outer portions. The masses of cloud which surround the planet seem to be immensely thick, and as clouds are, of course, light bodies in comparison with their bulk, they have the effect of largely increasing the apparent size of Jupiter, while adding very little to his weight. There is thus a great deal of mere inflation about this planet, by which he looks much bigger than his actual materials would warrant if he were constituted like the earth.

These facts suggest an interesting question. Why has Jupiter such immense surroundings? The clouds we are so familiar with down here on the earth are produced by the heat of the sun, which beats down upon the wide surface of the ocean, evaporates the water, and raises the vapour up to where it forms the clouds. Heat, therefore, is necessary for the formation of cloud; and with clouds so dense and so massive as those on Jupiter, more heat would apparently be necessary than is required for the moderate clouds on this earth. From whence is Jupiter to get this heat? Have we not seen that the great planet is far more distant from the sun than we are? In fact, the intensity of the sun's heat on Jupiter is not more than the twenty-fifth part of what we enjoy from the same source. We can hardly believe that the sun supplied the heat to make those big clouds on the great planet; so we must look for an additional source, which can only be inside the planet itself. So far as his internal heat is concerned, Jupiter seems to be in much the same condition now as our earth was once, ages ago, before its surface had cooled down to the present temperature. As Jupiter is so much larger than the earth, it has been slower in parting with its heat. The planet seems not yet to have had time to cool sufficiently to enable water to lie on his surface. Thus the internal heat of the planet supplies an explanation of its clouds. We may also remark that as the present condition of Jupiter illustrates the early condition of our earth, so the present condition of the earth foreshadows the future reserved for Jupiter when he shall have had time to cool down, and when the waters that now exist in the form of vapour shall be condensed into oceans on his surface.

THE SATELLITES OF JUPITER.

Every owner of a telescope delights to turn it on the planet Jupiter, both for the spectacle the globe itself affords him and for a view of the wonderful system of satellites by which the giant planet is attended. Fortunately, the satellites of Jupiter lie within reach of even the most modest telescope, and their incessant changes relative to Jupiter and each other give them a never-ending interest for the astronomer. Compared with the torpid performance of our moon, which requires a month to complete a circuit around the earth, Jupiter's moons are wonderfully brisk and lively. Nor are they small bodies like the satellites of Mars, for the second of Jupiter's satellites is quite as big as our moon, and the other three are very much larger. It is, however, true that his satellites appear insignificant when compared with Jupiter's own enormous bulk.

The innermost of these little bodies flies right round in a period of one day and eighteen or nineteen hours, while the outermost of them takes a little more than a fortnight—that is, rather more than half the time that our moon demands for a complete revolution. Jupiter's satellites are too far off for us to see much with respect to their structure or appearance even with mighty telescopes. It is, of course, their great distance from us that makes them look insignificant. They would, however, be bright enough to be seen with the naked eye, like small stars, were it not that, being so close to Jupiter, his overpowering brightness renders such faint objects in his vicinity invisible.

It was by means of the satellites of Jupiter that one of the most beautiful of scientific discoveries was made. As a satellite revolves round Jupiter it often happens that the little body enters into the shadow of the great planet. No sunlight will then fall upon the satellite; and as it has no light of its own, it disappears from sight until it has passed through the shadow and again receives sunlight on the other side. We can watch these eclipses with our telescopes, and there can be no more interesting employment for a small telescope. The movements of Jupiter's satellites are now known so thoroughly that the occurrence of the eclipse can be predicted. The almanacs will tell when the satellite is calculated to disappear, and when it ought to return again to visibility. When astronomers first began to make these calculations, a couple of hundred years ago, the little satellites gave a great deal of trouble. They would not keep their time. Sometimes they were a quarter of an hour too soon, and sometimes a quarter of an hour too late. At last, however, the reason for these irregularities was discovered, and a wonderful reason it was.

Suppose that there were a number of cannons all over Hyde Park, and that these cannons were fired at the same moment by electricity. Though the sounds would be all produced simultaneously, yet, no matter where you stood, you would not hear them all together; the noise from the cannons close at hand would reach your ears first, and the more distant reports would come in gradually afterwards. You can calculate the distance of a flash of lightning from you if you allow a mile for every five seconds that elapse between the time you saw the flash and the time you heard the peal of thunder which followed it. The light and the noise were produced simultaneously, but the sound takes five seconds to pass over every mile, while the light, in comparison to sound, may be said to move instantaneously. That sound travelled with a limited velocity was always obvious, but never until the discrepancies arose about Jupiter's satellites was it learned that light also takes time to travel. It is true that light travels much more quickly than sound—indeed, about a million times as fast. Light goes so quickly, that it would rush more than seven times around the earth in a single second. So far as terrestrial distances are concerned, the velocity of light is such that the time it requires for a journey is inappreciable. The distances, however, between one celestial body and another are so enormous, that even a ray of light, moving as quickly as it alone can move, will occupy a measurable time on the road. Our moon is so near us, that light takes little more than a second to cover that short distance.

Eight minutes are, however, required for light to journey from the sun to the earth; in fact, the sunbeams that now come into our eyes left the sun eight minutes ago. If the sun were to be suddenly extinguished, it would still seem to shine as brightly as ever in the eyes of the inhabitants of this earth for eight minutes longer. As Jupiter is five times as far from the sun as we are, it follows that the light from the sun to Jupiter will spend forty minutes on the journey, and conversely the light from Jupiter to the earth will take the same period. When we look at Jupiter and his moons, we do not see him as he is now, we see him as he was more than half an hour ago, but the interval will vary somewhat according to our different distances from the planet. Sometimes the light from Jupiter will reach us in as little as thirty-two minutes, while sometimes it will take as much as forty-eight — that is, the light sometimes requires for its journey a quarter of an hour more than is sufficient at other times.

We can, therefore, understand that irregularity of Jupiter's satellites which puzzled the early astronomers. An eclipse sometimes appeared a quarter of an hour before it was expected; because the earth was then as near as it could be to Jupiter, while the calculations had been made from observations when Jupiter was at his greatest distance. It was the eclipses of the satellites which first suggested the possibility that light must have a measurable speed. When this was taken into account, then the occasional delay of the eclipses were found to be satisfactorily explained. Confirmation flowed in from other sources, and thus the discovery of the velocity of light was completely established.

SATURN.

Next outside Jupiter on the confines of the ancient planetary system revolves another grand planet, called Saturn. His distance from us is sometimes nearly a thousand millions of miles, and he requires more than a quarter of a century for the completion of each revolution. Sometimes people do not pronounce the names of the planets

Fig. 66.—Saturn and Earth compared.

quite correctly. I have heard of a gardener who has a taste for astronomy, and sometimes begins to talk about the planets Juniper and Citron. Probably you will know what he meant to say. The ancients had discovered Saturn to be a planet, for though he looked like a star, yet his movement through the constellations could not escape their notice when attention was paid to the heavens.

In the matter of size Saturn is only surpassed by Jupiter among the planets. He is about 600 times as large as the earth; the small object, E, shown in Fig. 66 represents our earth in its true comparative size to the ringed planet; but Saturn is so far off, that even at his best he is never so bright as Venus, or Mars, or Jupiter become when they are favourably situated. On the globe of Saturn we can sometimes see a few bands, but they are faint compared with those on Jupiter. There is, however, no doubt that what we see upon Saturn is a dense mass of clouds. Indeed, he can have comparatively little solid matter inside, for this planet does not weigh so much as a ball of water the same size would do. Saturn, like Jupiter, must be highly heated in his interior.

The ring, or rather series of rings, by which the planet is surrounded are also shown in Fig. 66: these appendages are not fastened to the globe of Saturn by any material bonds, they are poised in space, without any support, while the globe or planet proper is placed symmetrically in the interior.

I have made a model which shows Saturn with his rings, but it is necessary for me to fasten the rings by little pieces of wire to the globe, for there is no mechanical means by which the rings of the model could be poised without support, as they are around the planet. If we throw the beam of the electric lamp on the little planet, we see the shadow which the planet casts on its ring. Similar shadows can be observed in the actual Saturn of the sky, and this is a proof that the planet does not shine by its own light, but by the light of the sun which falls upon it. Here again we illustrate the wide difference between a planet and a star, for were our sun to be put out, Saturn and all the other planets in the sky would vanish from sight, while the stars would, of course, twinkle on as before. There are three rings surrounding Saturn; they all lie in the same plane, and they are so thin, that when turned edgewise towards us the whole system almost disappears, except in very powerful telescopes. The outer and the inner bright rings are divided by a dark line, which can be traced entirely round. At the inner edge of the inner ring begins that strange structure called the *crape* ring, which extends half-way towards the globe of the planet. The most remarkable point about the crape ring is its semi-transparency, for we are sometimes permitted to see the globe of the planet right through this strange curtain. The crape ring can only be observed with a powerful telescope. The other two rings are within the power of very moderate instruments.

THE NATURE OF THE RINGS.

For the explanation of the nature of Saturn's rings we are indebted to the calculations of mathematicians. You might have thought, perhaps, that nothing would be simpler than to suppose the rings were stiff plates made from some solid material. But the question cannot be thus settled. We know that the weight of the ring could not be borne if it was a solid body. I may illustrate the argument by familiar facts about bridges. Where the span is but a small one, as, for instance, when a road has to cross a railway, a canal, or a river, the arch is, of course, the proper kind of structure. There is, for example, a specially beautiful arch over the river Dee at Chester. But if the bridge be longer than this, masonry arches are not suitable. Where a considerable span has to be crossed at the Menai Straits, or a gigantic one at the Firth of Forth, then arches have to be abandoned, and iron bridges of a totally different construction have to be employed. Arches cannot be used beyond a limited span because the strain upon the materials becomes too great for their powers of resistance to withstand. Each of the stones in an arch is squeezed by intense pressure, and there is a limit beyond which even the stoutest stones cannot be relied upon. As soon, therefore, as the span of the arch is so great that the stones it contains are squeezed as far as is compatible with safety, then the limit of size for that form of arch has been reached.

Suppose that you stood on Saturn at his equator, and looked up at the mighty ring which would stretch edgewise across your sky. It would rise up from the horizon on one side, and, passing over your head, would slope down to the horizon on the other. You would, in fact, be under an arch of which the span was about 100,000 miles. Owing to the attraction of Saturn, every part of that structure would be pulled forcibly towards his surface, and thus the materials of the arch, if it were a solid body, would be compressed with terrific force.

It does not really matter that the arch I am now speaking of is half of a ring the other half of which is below the globe of the planet. That is only a difference with respect to the supports of the two ends of the arch, and does not affect the question as to the pressure upon its materials; nor does the fact that the ring is revolving remove the difficulty, though it undoubtedly lessens it. We know no solid substance which could endure the pressure. Even the toughest steel that ever was made would bend up like dough under such influence. We cannot, therefore, account for Saturn's ring by supposing it to be a solid, for no solid would be strong enough.

Do not you remember the old fable of the oak tree and the pliant reed, how when the storm was about to arise the oak laughed at the poor reed, and said it would never be able to withstand the blasts? But matters did not so turn out. The mighty oak, which would not yield to the storm, was blown down; while the little reed bent to the wind and suffered no injury. This gives us a hint as to the true constitution of Saturn's ring; it is not a solid body, trying to resist by mere strength: it is rather to be explained as an excessively pliant structure. Indeed, I ought not to call it a structure at all; it is rather a multitude of small bodies not in the least attached together. I do not know what the size of these bodies is. For anything we can tell, they may be no larger than the pebbles you see on a gravel walk.

Let us see how we could encircle our earth with rings like those which surround Saturn. I shall ask to be provided with a sufficiently large number of pebbles, and you must also imagine that I have the means of ascending high up into space, half-way from here to the moon. Suppose I went up there and simply dropped the pebble, of course it would tumble straight down to the earth again. If, however, I threw it out with proper speed and in the proper direction, I could start it off like a little moon, and it would go on round and round our earth in a circle. It is no matter what the size of the pebble is—it may be as small as a grain of shot or as big as a cannon-ball. Then take another pebble. Cast it also in a somewhat similar path, taking care, however, that the planes of the two orbits shall be the same. Each of these little bodies will pursue its journey without interference from the other. Then proceed in the same way with a third, a fourth, with thousands and millions and billions of pebbles, until at last the little bodies will become so numerous, that they almost fill a large part of the plane with a continuous shoal. Each little object, guided entirely by the earth's attraction, will pursue its path with undeviating regularity. Its neighbours will not interfere with it, nor will it interfere with them. Let us circumscribe the limits of our flat shoal of moonlets. We first take away all those that lie outside a certain large circle; then we will clear away sufficient to make a vacant space between the outer ring and the inner ring, and thus the two conspicuous rings have been made; at the inside of the inner ring we will pick out numbers of the pebbles here and there, so as to make this part much less dense than the outer portions, and thus produce a crape ring; then we will clear away those that come too close to the planet, and form a neat inner boundary.

Could we then view our handiwork from the standpoint of another planet, what appearance would our earth present? The several pebbles, though individually so small, would yet by their countless numbers reflect the sun's light so as to produce the appearance of a continuous sheet. Thus we should find a large bright outer ring surrounding the earth, separated by a dark interval from the inner ring, and then from the margin of the inner ring the pebbles would be so much more openly distributed that we should be able to see through them to some extent. That beautiful system of rings which Saturn displays to us is undoubtedly of a similar character to the fictitious system of rings which I have endeavoured to describe. No other explanation will account for the facts, especially for the semi-transparency of the crape ring. The separate bodies from which Saturn's rings are constituted seem, however, so small that we are not able to see them individually. There are some other fine lines running round the rings beside the great division, and these are also explained by the theory that I have stated.

Saturn has other claims on our attention besides those of its rings. It has an elaborate retinue of satellites—no fewer, indeed, than eight; but some of them are very faint objects, and not by any means so interesting as the system by which Jupiter is attended.

Saturn was the last and outermost of the planets with which the ancients were acquainted. Its path lay on the frontiers of the then known solar system, and the magnificence of the planet itself, with its attendant luminaries and its marvellous rings, rendered it worthy indeed of a position so dignified. These five planets—namely, Mercury, Venus, Mars, Jupiter, and Saturn—made up with the sun and moon the seven "planets" of the ancients. They were supposed to complete the solar system, and, indeed, the existence of other members was thought to be impossible. In modern times it has been discovered that there are yet two more planets. I do not now refer to those little bodies which run about in scores between Mars and Jupiter. I mean two grand first-class planets, far bigger than our earth. One of them is Uranus, which revolves far outside Saturn, and the other is Neptune, which is much further still, and whose mighty orbit includes the whole planetary system in its circuit. To complete this journey not less than 165 years is required.

WILLIAM HERSCHEL.

I have to begin the account of this discovery by telling you a little story. In the middle of the last century there lived at Hanover a teacher of music whose name was Isaac Herschel. He had a large family of ten children, and he did the best for them that his scanty means would permit. Of his children William was the fourth, and he inherited his father's talents for music, as did most of his brothers and sisters. He was a bright, clever boy at school, and he made such good progress in his music that by the time he was fourteen years old he was able to play in the military band of the Hanoverian Guards. War broke out between France and England, and as Hanover was then under the English crown, the French invaded it, and a battle was fought in which the poor Hanoverian Guards suffered very terribly. Young Herschel spent the night after the battle in a ditch, and he came to the conclusion that he did not like fighting, even so far as being in the band was concerned, and he resolved to change his profession. That was not so easy to do just then, for even a bandsman cannot leave the service in war-time at his own free will. William Herschel, however, showed all through his life that he was not the man to be baffled by difficulties. I do not know whether he asked for leave, but at all events he took it. He deserted, in fact, and his friends succeeded in sending him away to England.

He was nineteen years old when he commenced to look for a career over here, and certainly at first he found his prospects in the musical profession very discouraging. Herschel was, however, very industrious; and at last he succeeded in getting appointed as organist of the Octagon

Chapel at Bath. He gradually became famous for his musical skill, and had numbers of pupils. He used also to conduct concerts and oratorios, and was well known in this way over the West of England. Busy as Herschel was with his profession, he still retained his love of reading and study. Every moment that he could spare from his duties he devoted to his books. It was natural that a musician should specially desire to study the theory of music. This is a difficult subject, and to understand it properly you should know Euclid and algebra, and, indeed, higher branches of mathematics as well. Herschel did not know these things at first; he had not the means of learning them when he was a boy, so he worked very hard after he became a man. And he studied with such success that he made fair progress in mathematics, and then it appeared to him that it would be interesting to learn something about astronomy. After he had begun to read about the stars, he thought he would like to see them, and so he borrowed a telescope. It was only a little instrument, but it delighted him so much that he said he must have one for himself. So he wrote to London to make inquiries.

Telescopes were much dearer in those days than they are now, and Herschel could not give the price that the opticians asked. Here again his invincible determination came to his aid. What was there to prevent him from making a telescope? he asked himself; and forthwith he began the attempt. You will think it strange, perhaps, that a music-teacher who had no special training as a mechanic should at once commence so delicate and difficult a task; but it is not really so hard to make a telescope as might be imagined. The amateur cannot make so pretty-looking an instrument as he will be able to buy at the shops—the tubes will not be so smart, and the finish will be such as a trained workman would be ashamed of—but the essential part of a telescope is comparatively easy to make; at least, I should say of a reflecting telescope, which is the kind Herschel attempted to make, and succeeded in making. You must know that there are two kinds of telescopes. The more ordinary one that you are all familiar with is called the refracting telescope, and it has glass lenses. It was a telescope on this principle that we spoke of in a former lecture (p. 92). The reflecting telescope depends for its power upon a bright mirror at the lower end, and when using this telescope you look at the reflection of the stars in this mirror. It was a reflector like this that Herschel began to construct, and he engaged in the task with enthusiasm. His sister Caroline had come to live with him, and she used to help him at his work. So much in earnest was he that he used to rush into his workshop directly he came home from a concert, and without taking off his best clothes he would plunge again into the grinding and polishing of his mirrors. His sister tried to keep the house as tidy as possible, but Herschel put up a carpenter's shop in the drawing-room, and turning lathes in the best bedroom. At last he succeeded. He made a mirror of the right shape, and found that it exhibited the stars properly. It was not a mirror in the ordinary sense, with glass on one side and quicksilver on the other. The mirror that Herschel constructed was entirely of metal. It consisted of a mixture of two parts of copper with one of tin.

The copper has first to be melted in a furnace, for the metal must be above a red heat before it will begin to run.

Then the tin has to be carefully added, and the casting of the mirror is effected by pouring the molten metal into a flat mould. Thus the rough mirror is obtained, which in Herschel's earlier telescopes seems to have been about six or seven inches in diameter, and nearly an inch thick. Though copper is such a tough substance, and though tin is also tough, yet when melted together to make speculum metal, as this mixture is called, they produce an exceedingly hard and brittle material. When we remember that we could never break a copper penny piece by throwing it down on the flags, it may seem strange that the "speculum metal"

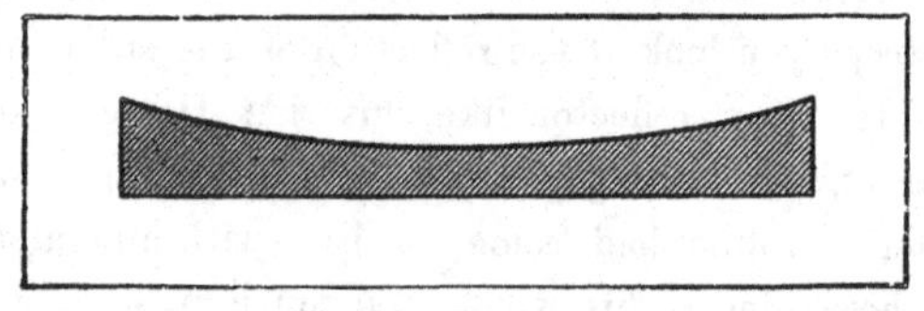

Fig. 67.—The Mirror.

should be so exceedingly brittle. A piece the size of a penny would be even more brittle than a bit of glass of the same dimensions, and when the casting is made, unless it is cooled very carefully, it will fly in pieces. Herein lay one of the difficulties that Herschel encountered. Speculum metal must be put into an oven as soon as the casting has become solid, and then the heat is gradually allowed to abate. When the disc has been at last obtained, next follows the labour of giving it the true figure and polish. It is not only more fragile than glass, but it is also quite as hard as glass, and therefore the grinding is a tedious operation. First the surface has to be ground with coarse sand, and then with emery, which has gradually to be finer and finer until the true figure has been given (Fig. 67). The mirror is then slightly basin-shaped, but the depression is very little. For example, in a mirror six inches across the depression at the centre would perhaps be not more than the twentieth of an inch. Small though this depression is, yet it has to be given with exactness. In fact, if it were wrong at any point by so much as the tenth of the thickness of this sheet of paper, the telescope would not perform accurately. The tool that is used in grinding is made of iron, and has been turned in a lathe to the right shape. It is divided into squares, in the manner shown in Fig. 68. After the grinding is over comes the polishing, and this is effected with a tool like the grinder in shape. This has to be covered over with little squares of pitch, so that when warmed and put down on the mirror it is soft enough to receive the right shape. A little rouge and water is spread over the mirror, and the polisher is worked backward and forward with the hand until a brilliant surface is obtained.

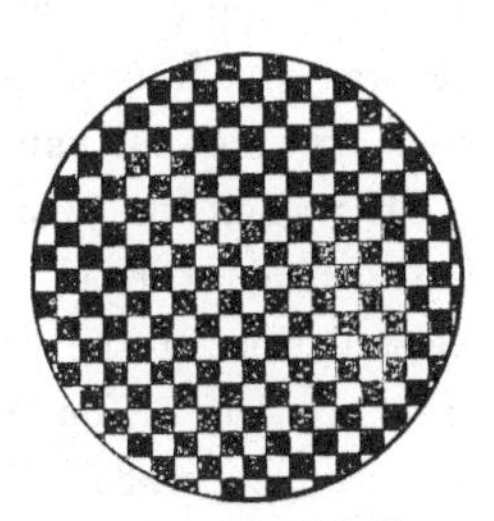

Fig. 68.—The Grinding Tool.

When the amateur astronomer has completed this part of the task, all the great difficulties about his telescope are conquered. The tube may be made of wood, and, indeed, a square tube will do just as well as a round one. He must also provide for the top of the tube a small mirror, which has to be perfectly flat. The preparation of this requires much care, because it is not so easy as one might suppose to obtain an accurately flat surface. You must get three pieces, and grind each two of them together until each pair will touch all over; then they will certainly all be flat. One more part you want, and that is an eye-piece. This presents no difficulty. A single glass lens can be made to answer, and your telescope is complete.

THE DISCOVERY OF URANUS.

It is now more than 100 years (1774) since Herschel first had a view of the heavens through the telescope he had himself constructed. For the first few years he does not appear to have made any important discoveries. He was gradually preparing himself for the great achievement by which his name became famous.

It was on the 13th of March, 1781, that the organist of the Octagon Chapel at Bath turned his telescope on the constellation of the Twins, and began to look at one star after another. You must know that a star merely looks like a little point in a telescope; even the greatest instrument will only make the star look brighter, and will never show it with a perceptible disc. In looking over the stars this night, Herschel's attention was arrested by one object that did look larger when magnified, and therefore was not a star. The only other objects which would behave in this way were the planets or possibly a comet. Indeed, at first Herschel imagined that what he saw must be a comet. It could hardly have occurred to him that he would have such good fortune as to discover a new planet. The five great planets had been known from all antiquity. Was it reasonable to suppose that there could be yet another that had never been perceived? Fortunately, there was a test available. A star remains in the same place from night to night and from year to year; while a planet, as we have already had occasion to mention, is a body which is wandering about. The movements are, however, not at all like those of a comet. To decide on the nature of Herschel's newly discovered body, it was sufficient to observe the character of its motion. A few nights sufficed to do this. The position of the body was carefully marked relatively to the neighbouring stars, and it was soon shown that it was a planet.

Here, then, a great discovery was made. A new planet, now called Uranus, was added to our system. It would be nothing to discover a new star. You might as well talk of discovering a new grain of sand on the sea-shore. The stars are in untold myriads. They are so far off that they have no relation whatever to our system, which is presided over by the sun. But by the detection of a new planet, revolving far outside Saturn, Herschel showed that a new and most interesting member had to be added to the five old planets which have been known from the earliest records of history.

It may well be imagined that a discovery so startling as this excited astonishment throughout the scientific world. "Who is this Bath organist?" everybody asked. Accounts of him and his discoveries appeared in the papers. They were not then as familiar with the name of Herschel as we are, happily, now; and the spelling of the unusual name showed many varieties. When George III. heard of Herschel's great achievement, he directed the astronomer to be summoned to Windsor, that the King might receive an account of the wonderful discovery from the lips of the discoverer himself. Herschel of course obeyed, and he brought with him his famous telescope, and also a map of the whole solar system, to show to the King. No doubt he thought that his Majesty had probably not paid much attention to astronomy. Herschel was, therefore, prepared to explain to the King what it would be necessary for him to know before he would be fully able to appreciate the magnitude of the discovery.

You will remember that Herschel while still a boy had deserted from the army, many years previously. It appears that the King had learned this fact in some way, so that when Herschel was ushered into his presence his Majesty said that before the great astronomer could discuss science there was a little matter of business that must be disposed of. The King accordingly handed Herschel a paper, in which he was, I dare say, greatly surprised to find a pardon to the deserter written out by the King himself.

Then Herschel unfolded his wonderful discovery, which the King thoroughly appreciated, and in the evening the telescope was called into action in the gardens, and the glories of the heavens were displayed. Herschel made a most favourable impression on his Majesty, and when the King told the ladies of the Castle next day of all that Herschel had shown him, their astronomical ardour was also aroused, and they asked to see through the marvellous tube. Of course Herschel was ready to comply, and the telescope was accordingly carried to the windows of the Queen's apartments at Windsor, which commanded a fine view of the sky; I should rather say, they would have commanded a fine view if the clouds had not been in the way, which they unfortunately were. Even for royalty the clouds would not disperse, so what was to be done? Herschel was equal to the occasion. He specially wanted to exhibit Saturn, for it is one of the most beautiful objects in the sky, and will fascinate any intelligent beholder. Other astronomers would not have been able to see Saturn through the clouds, but Herschel did not disappoint his visitors; he directed the instrument, not to the sky (nothing was there to be seen): he turned it towards a distant garden wall. Now what would you expect to see by looking through a telescope at a garden wall—bricks, perhaps, or ivy? What these ladies saw was a beautiful image of Saturn, his globe in the centre and his rings all complete, forming so true a resemblance to the planet that even an experienced astronomer might have been deceived. In the afternoon Herschel had seen that the clouds were thick, and that there would be little probability of using the telescope properly. Accordingly, he cut out a little image of Saturn, illuminated it by lamps, and fixed it on the garden wall.

Herschel's visit to Windsor was productive of important consequences. The King said it was a pity that so great an astronomer should devote himself to music, and that it would be far better for him to give up that profession and come and live at Windsor. His Majesty promised that he would pay him a salary, and he also undertook to provide the cost of erecting great telescopes. His faithful sister Caroline came with him as his assistant, and also received some bounty from the King. From that moment Herschel renounced all his musical business; and devoted himself to his great life-task of observing the heavens.

He built telescopes of proportions far exceeding those that had ever been then thought of. He used to stand at night in the open air from dusk to dawn gazing down the tube of his mighty reflector, watching the stars and other objects in the heavens as they moved past. He would dictate what he saw to Caroline, who sat near him. It was her business to write down his notes and to record the position of the objects which he was describing. Sometimes, she tells us, the cold was so great that the ink used to freeze in her pen when she was at this work. Until he became a very old man, Herschel devoted himself to his astronomical labours. His discoveries are to be counted by thousands, though not one of them was so striking or so important as the detection of the new planet which first brought him fame.

The question of a name for the addition to the sun's family had, of course, to be settled. Herschel had surely a right to be heard at the christening, and as a compliment to his Majesty he named the stranger the Georgium Sidus. So, indeed, for a brief while, the planet was actually styled. The Continental astronomers, however, would not accept this designation; all the other planets were named after ancient divinities, and it was thought that the King of England would seem oddly associated with Jupiter and Saturn; perhaps also they considered that the British dominions, on which the sun never sets, were already quite large enough, without further extension to the celestial regions. Accordingly a consultation was held, the result of which was that George III. was dethroned from planetary honours, and the body was given the name of Uranus, which, by universal consent, it now bears.

The planet Uranus lies just on the verge of visibility with the unaided eye. It can sometimes be glimpsed like a faint star, and, of course, with a telescope it is readily perceived. Many generations of astronomers before Herschel's time had been observing the heavens, making maps of the stars, and compiling great catalogues in which the places of the stars were accurately put down. It often happened that Uranus came under their notice, but it never occurred to them that what seemed so like a star was really a planet. I have no doubt said that Uranus looked unlike a star when Herschel examined it; but then that was because Herschel was a particularly skilful astronomer. To an observer of a more ordinary type Uranus would not present any very remarkable appearance, and would be passed over merely as a small star. In fact, the planet was thus observed not once or twice, but no fewer than seventeen times before the acute eye of Herschel perceived its true character. On many of the previous occasions the planet had been noted as a star by astronomers who are in every way entitled to our respect. It required a Herschel, determined to see everything in the very best manner, to grasp the discovery which eluded so many others.

When Uranus was observed on these former occasions and mistaken for a star, its place was carefully put down. These records are at present of the utmost use, because they show the past history of the planet; and they appear all the more valuable when we remember that Uranus requires no less than eighty-four years to accomplish a single revolution around the sun. Thus, since the planet was discovered in 1781, it had completed one revolution by 1865, and is now (1889) about one-third of the way around another. The earlier observations extend backwards almost 200 years, so that altogether we have more or less information about the movements of the planet during the completion of two circuits and a half.

Uranus is a great deal bigger than the earth, as you will see in the view of the comparative sizes of the planets (Fig. 46). It appears to be of a bluish hue, but we cannot tell whether it turns round on its axis, or rather, I should say, we cannot *see* whether it turns round on its axis; for we can hardly doubt that it does so. Uranus also seems to be greatly swollen by clouds, in the same manner as are both Jupiter and Saturn; in fact, if our earth were as big as Uranus, it would weigh four or five times as much as Uranus does. Hence we are certain that Uranus must consist of materials less dense on the whole than are those of which our earth is made.

THE SATELLITES OF URANUS.

You must use a very good telescope to see the satellites of Uranus. They are four in number, bearing the names of Ariel, Umbriel, Titania, and Oberon. The innermost of these, Ariel, completes a journey round the planet in two days and a half; Oberon, the most distant, requires thirteen days and a half.

Notwithstanding that Uranus is at so great a distance from the earth, we have been able to put this planet, as well as the nearer ones, in the weighing scales, and we assert with confidence that Uranus is fifteen times as heavy as our earth. We are indebted to the satellites for this information. A planet is always tending to pull its satellite down, and the satellite is kept from falling by the speed with which it revolves. The heavier the planet, the faster do its satellites go round. Thus, to take an illustration from our own moon, we know that, if the earth were to be made four times heavier than it is, the moon would have to spin round twice as fast as it does, in order to remain in the same orbit. The speed with which the satellites of Uranus revolve accordingly affords a measure of the mass of the planet. Were Uranus heavier than he is, his satellites would revolve more quickly than they do; were he lighter, the satellites would take a longer period for their revolutions.

There is another singular circumstance connected with the moons of Uranus. I have told you how every body revolving around another by gravitation will describe an ellipse; but, of course, there are many different kinds of ellipses, and some may be nearly circles. There is nothing whatever to prevent a satellite from revolving around its primary in an exact circle if it be started properly; that is, in the right direction and with the right speed. Most of the motions in the planetary system are not exactly circular. A notable exception is found in the case of Uranus. All the four satellites of this planet seem to revolve in circles so perfect that we can make a true picture of this system with a pair of compasses. It is further to be noticed that the four circles seem to lie exactly in the same plane. In general the orbits of the great planets and of their satellites lie very nearly in the ecliptic in which the earth moves. Here again the satellites of Uranus are exceptional. The plane in which they are contained stands up almost squarely to the plane in which the motion of the planet is performed. The moons of Uranus seem to have got a twist, from some accidental circumstance for which we are not able to account.

THE DISCOVERY OF NEPTUNE.

The boundaries of the solar system had been much extended by the discovery of Uranus, but they were destined to receive still further enlargement by the detection of another vast planet, revolving far outside Uranus, the orbit of which forms, according to our present knowledge, the outline of the planetary system.

I have here to describe one of the greatest discoveries that have ever been made. It is not the magnificence of the outermost planet itself that I refer to, though, indeed, it is bigger than Uranus. I am rather thinking of the *way* in which the discovery was made. I do not mean any disrespect to Herschel when I say that the discovery of Uranus was chiefly a stroke of good fortune; but I may be permitted to describe it in this manner by way of emphasising as strongly as I can how utterly different was the train of ideas which led to the discovery of Neptune. Herschel merely looked at one star after another till suddenly he dropped on the planet, having beforehand not the slightest notion that any such planet was likely to exist. But Neptune was shown to exist before it was ever seen, and, in fact, the man that first saw the planet, and knew it to be a planet, was not the discoverer. This is rather a difficult subject; and it would take you years of hard study to be able to follow the train of reasoning by which Neptune was found. I shall, however, make an attempt to explain this matter sufficiently to give at least some idea of the kind of problem that had to be solved.

You will remember that law of Kepler which tells us that every planet moves round the sun in an ellipse. If the planet be uninterfered with in any way and guided only by the attraction of the sun, it will for ever continue to describe precisely the same ellipse without the slightest alteration. It was ascertained that the path which Uranus followed was not always regular. The early observations of the planet, when it was mistaken for a star, have here been of the utmost service. They have indicated the ellipse which Uranus described the last time it went round, and our own modern observations have taught us the path which the planet is at present describing. These two ellipses are slightly different, and the consequence is that, supposing we take the observations of Uranus made 100 years ago, and calculate from them where Uranus ought to be now, we find that the planet is a little astray. Astronomers are not accustomed to be wrong in such calculations, and when there are discrepancies the first thing to be done is to see what has caused them. It is certain that the position in which Uranus is found this very night, for example, is not what it would have been had the sun alone been guiding the planet. Perhaps you will think that it is impossible for reliable calculations to be made about such matters; but I assure you they can, and the very fact that Uranus was a little astray made it interesting to try and find out the cause of the disturbance.

I have already explained when speaking about Mars, (p. 176), that there is an attraction between every two bodies, but in the group of planets to which the earth belongs the sun's attraction is so much stronger than any other force that all the movements are mainly guided by it. Nevertheless it is true that not only does the sun pull our earth and all the other planets as well, but all the planets, including the earth, are pulling one another. In fact, there is an incessant struggle going on in the family party. Fortunately the sun is so much more powerful than any other member, that he keeps them all pretty well in order; and unless you look very carefully you will not see the effects of the little struggles that are going on between every pair of the system. Our earth itself is pulled and swayed to and fro by the actions of its brothers and sisters. It is dragged perhaps a thousand or two thousand miles this way by Jupiter, or it gets a good tug in the other direction by Venus. Mars and Saturn also do their little best to force the earth away from its strict path. However, our earth does not suffer much from these irregularities. It pursues its route fairly enough, just as a coach from London to Brighton will get safely to its destination notwithstanding the fact that it has to swerve a little from its path whenever it meets other vehicles on the way, or when the coachman wishes to avoid a piece of the road on which stones have been freshly laid down.

The track followed by Uranus was found to be somewhat erratic, like that of every other planet. Jupiter gave it a pull, and so did Saturn, and at first it was thought that the irregularities which were perceived could be explained by the action of these planets, so big and so well known. Here is a question for calculation; it involves a very long and a very hard piece of work, but it is possible to estimate how far each of the other planets is capable of dragging Uranus from its path. It is remarkable that by working out sums we should be able to find what one planet hundreds of millions of miles away was able to do to another planet that was further still, and that not alone for to-day or for yesterday, but for past time extending over more than a century. If, however, you will listen to me a little longer, I think I shall give you a proof that these sums could be worked out correctly.

When the calculations had been made which showed how much the known planets could disturb Uranus, it was found that there were still some deviations of the planet that remained unexplained. They were not large; they only amounted to showing that the body was just a little astray from the spot where the calculations indicated it should be. The rest of astronomy was so perfect, and the law of attraction prevailed so universally, that it was thought the law of attraction must provide some way of explaining the behaviour of Uranus. He could not have left his track of his own accord; therefore there must be some agency at work upon him of which we did not know. What could be this unknown source of disturbance? Every such trouble had hitherto been found to be a consequence of the attraction of gravitation; therefore there must be some body pulling at Uranus which no one has ever seen. Where could it be? How was it to be discovered? Such were the questions that were asked, and they were answered in a most satisfactory manner.

First of all, what sort of body could it be that was pulling Uranus? It is obvious that none of the stars would be competent to produce so great an effect; they are all so far off that they have nothing whatever to say to any of the domestic matters in our little solar system, which is simply a group by itself. It would be more reasonable to suppose that there must be another planet which nobody had ever seen, but which affected Uranus so as to account for his truant behaviour. To begin to search for this planet with telescopes without some guidance would be futile; in fact, astronomers had been searching the heavens for planets for nearly fifty years, and though several had been discovered, they all belonged to the zone of little planets, and none of them were big enough to pull about Uranus appreciably. Of course, if all the stars could be blotted out of the sky, so that nothing but planets were left, then, by sweeping the telescope over the heavens, every planet that exists might be speedily picked up. The difficulty is that the planets, which are either small or very distant, look so like the stars that it is impossible to recognise them among the millions of glittering points in the sky. It was, however, hoped that the unknown planet would be large enough to be visible in the telescope, if only we knew exactly where to point it..

Two illustrious astronomers, Adams of Cambridge, and Leverrier of Paris, both separately undertook an astonishing piece of calculation. They tried to find out where the unknown planet must be from the mere fact that it deranged Uranus in a particular way. I daresay many of those who are reading this book have learned simple equations in algebra, and they have worked such questions as to find the length of a pole half of which is in mud, a quarter in water, and ten feet above the water. Those who know this much can perhaps realise the problem that had to be solved in trying to discover the unknown planet. So difficult a question as this had to be solved in a way that your masters would hardly allow you to use when working your sums in algebra. I do not think they would let you make a series of guesses. Let us try 20 feet, for instance, as the length of the pole; that will make 10 feet in the mud, 5 feet in the water, and 5 feet outside. This will not do; it is not long enough; we must try again; and after another guess or two, we see that a pole 40 feet long will exactly answer. We do not use this method of guessing in algebra, because solving the simple equation is a much better method. Adams and Leverrier found that to discover the unknown planet was a question so very difficult,

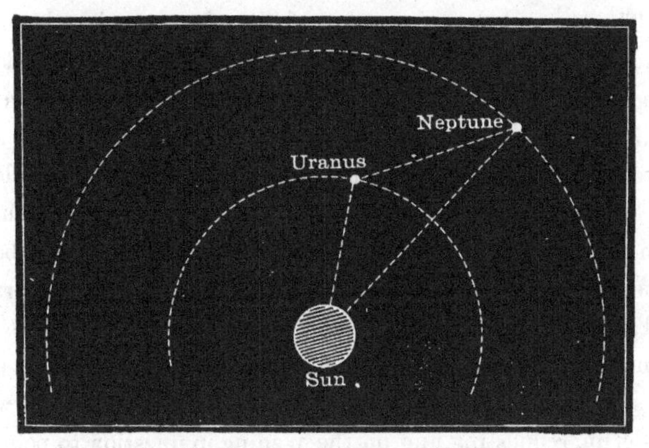

Fig. 69.—Orbits of Uranus and Neptune.

that they were obliged to use a sort of guessing, but very intelligent guessing, I need hardly assure you. They proceeded in this way (Fig. 69). They would draw a circle outside Uranus, and then they would suppose that a planet was revolving in that circle. They would calculate its effect upon Uranus, and see if it would account for what was observed. The first planet they tried would not do; then they began again with another, until at last, after many trials and much very hard work, they began to see that there might be a planet in a particular path far outside Uranus, such that if this planet were of the right weight and moving with the right speed, then it would pull Uranus exactly in the way that astronomers had observed it to be pulled. They found at last that there could be little doubt about the matter; for this unknown body would account for all the facts. Then, indeed, they had solved their equation; they had found the unknown; they had shown that it would do what was wanted.

The two great astronomers had thus discovered a planet, but as yet it was only a planet on paper. Those who could judge of the subject had no doubt that the planet was really in the sky; but just as you like to prove that you have found the correct answer to your sum, so people were naturally anxious to prove the truth of this wonderful sum that Adams and Leverrier had solved. This was to be done by actually seeing the planet of which the astronomers had asserted the existence. Leverrier calculated that the new planet on a certain night would be in a particular position on the sky. Accordingly he wrote to Dr. Galle, of the observatory at Berlin, requesting him on the evening in question to point his telescope to the very spot indicated, and there he would see a planet which human eyes had never before beheld. Of course, Dr. Galle was only too delighted to undertake so marvellous a commission. The evening was fine; the telescope was opened; it was directed towards the heavens; and there, in the very spot which the calculations of Leverrier had indicated, shone the beautiful little planet. At Cambridge arrangements had also been made to search for the new member of the solar system, in accordance with Professor Adams' calculations. There also the planet that had given all this trouble to Uranus was brought to light.

At first it looked like a star, as all such planets do; but that it was not a star was speedily proved, by the two tests which are infallible indications of a planet. First the body was so moving that its position with respect to the adjacent stars was constantly changing. Then, when a strong magnifying power was placed on the telescope, the little object was seen, not to be a mere starlike point, but to expand into the little disc which shows us we are not looking at a distant sun, but at a world like our own.

Was not this truly a great discovery? Have we not shown you how entitled the calculations of astronomers are to our respect, when we find that they actually discovered the existence of a majestic planet before the telescope had revealed it? See also the greatly increased interest that belongs to Herschel's discovery of Uranus. We can hardly imagine anything that would have given more gratification to this old astronomer than to think that his Uranus should have given rise to a discovery even more splendid than his own. He died, however, more than twenty years before this achievement.

The authorities who decide on such matters, christened the new planet Neptune; and this body wanders round on the outskirts of our solar system, requiring for each journey a period of no less than 165 years. The circle thus described has a radius thirty times as great as that of the earth's track.

Neptune is altogether invisible to the unaided eye, but it is sufficiently bright to have been occasionally recorded as a star. Indeed, nearly fifty years before it was actually discovered to be a planet it had been included by the astronomer Lalande in a list of stars he was observing. A curious circumstance was afterwards brought to notice. When reference was made to the books in which Lalande's observations were written, it was found that he had observed this star twice, namely, on May 8th and May 10th, 1785. Of course, if the object had been a star its position on the two days would have been the same, but being a planet it had moved. When Lalande, on looking over his papers, saw that the places of this supposed star were different on the two nights, he concluded that he must have made a mistake on the first night, and accordingly treated the object as if the place on the 10th was the right one. Just think how narrowly Lalande missed making a discovery! Unhappily for his renown, he took it for granted that one or both of his observations were erroneous, and so they must have been if the object had been a star. But they were both right; it was the planet which had moved in the interval.

As Neptune is half as far again from the earth as Uranus, we can hardly expect to learn much about the actual nature of the planet. We do know that it has four times the diameter of the earth, so that it exceeds the earth in the same proportion that the earth is larger than the moon.

Like the other great planets, Neptune is also enveloped with copious clouds; in fact, it only weighs one-fifth part as much as it would do if it were made of materials so substantial as are those of the earth. Like our earth, Neptune is attended by one moon, which revolves round the planet in a little more than six days.

And now our description has reached the boundary of our known system of planets. The five great planets of antiquity have been increased in these modern days by the addition of two more, Uranus and Neptune, while the discovery of a multitude of small planets has given a further increase to the number of the sun's family. We have still some other objects in our solar system to describe; some of them are excessively big: these are the comets. Some of them are exceedingly small: these are the shooting stars. We shall talk about these in our next lecture.

LECTURE V.

COMETS AND SHOOTING STARS.

The Movements of a Comet—Encke's Comet—The Great Comet of Halley—How the Telegraph is Used for Comets—The Parabola—The Materials of a Comet—Meteors—What Becomes of the Shooting Stars—Grand Meteors—The Great November Showers—Other Great Showers—Meteorites.

THE MOVEMENTS OF A COMET.

THE planets are all massive globes, more or less flattened at the poles; but now we have to talk about a multitude of objects of the most irregular shapes, and of the most flimsy description. We call them *comets*, and they exist in such numbers that an old astronomer has said "there were more comets in the sky than fishes in the sea," though I think we cannot quite believe him. There is also another wide difference between planets and comets: planets move round in nearly circular ellipses, and not only do we know where a planet is to-night, but we know where it was a month ago, or a hundred years ago, or where it will be a hundred years or a thousand years to come. All such movements are conducted with conspicuous regularity and order; but now we are to speak of bodies which generally come in upon us in the most uncertain and irregular fashion. They visit us we hardly know from whence, except that it is from outer space, and they are adorned in a glittering raiment, almost spiritual in its texture. They are always changing their appearance in a baffling, but still very fascinating manner. If an artist tries to draw a comet, he will hardly have finished his picture of it in one charming robe before he finds it arrayed in another. The astronomer has also his complaints to make against the comets. I have told you how thoroughly we can rely on the movements of the planets, but comets often play sad pranks with our calculations. They sometimes take the astronomers by surprise, and blaze out with their long tails just when we do not expect them. Then by way of compensation they frequently disappoint us by not appearing when they have been most anxiously looked for.

After a voyage through space the comet at length begins to draw in towards the central parts of our system, and as it approaches the sun, its pace becomes gradually greater and greater; in fact, as the body sweeps round the sun the speed is sometimes 20,000 times faster than an express train. It is sometimes more than 1,000 times as fast as the swiftest of rifle bullets, or at the rate of 200 miles a second. The closer the comet goes to the sun, the faster it moves; and a case has been known in which a comet, after coming in for an incalculable length of time towards the sun, has acquired a speed so tremendous, that in two hours it has whirled round the sun and has commenced to return to the depths of outer space. This terrific outburst of speed does not last long. A pace which near the sun is 20,000 times that of our express trains diminishes to 10,000, to fifty, to one; while in the outermost part of its path the comet seems to creep along so slowly that we might think it had been fatigued by its previous exertions.

We have so often seen a stream of sparks stretching out along the track of a sky-rocket, that we might naturally suppose the tail of a comet streamed out along its path in a somewhat similar manner. This would be quite wrong. You see from Fig. 70 that the tail does not

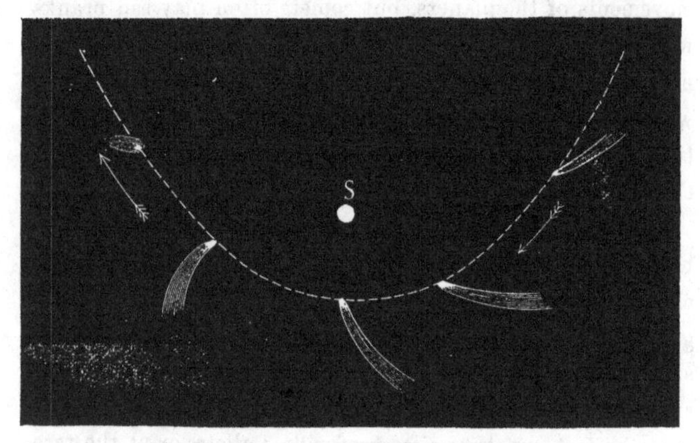

Fig. 70.—How the Comet's Tail is Disposed.

lie along the comet's path, but is always directed outwards from the sun. If you will draw a line from the sun to the head of the comet and follow the direction of the line, it shows the way in which the tail is arranged. You will also notice how the tail of the comet seems to grow in length as it approaches the sun. When the comet is first seen, the tail is often a very insignificant affair, but with enormous rapidity it shoots out until it becomes many millions of miles long, by the time the comet is whirling round the sun. These glories soon begin to wane as the comet flies outward; the tail gradually vanishes, and the comet retreats again to the depths of space in the same undecorated condition as that in which it first approached.

When a comet appears, it is always a matter of interest to see whether it is an entirely new object, or whether it may not be only another return of a comet which has paid us one or more previous visits. The question then arises as to how comets are to be identified. Here we see a wide contrast between unsubstantial bodies like comets and the weighty and stately planets. Sketches of the various planets or of the face of the sun, though they might show slight differences from time to time, are still always sufficiently characteristic, just as a photographic portrait will identify the individual, even though the lapse of years will bring some changes in his appearance. But the drawing of a comet is almost useless for identification. You might as well try to identify a cloud or a puff of smoke by making a picture of it. Make a drawing of a comet at one appearance, and sketch particularly the ample tail with which it is provided. The next time the comet comes round it may very possibly have two tails, or possibly no tail at all. We are therefore unable to place any reliance on the comet's personal appearance in our efforts to identify it. The highway which it follows through the sky affords the only means of recognition, for the comet, if undisturbed by other objects, will never change its actual orbit. But even this mark often fails, for it not unfrequently happens that during its erratic movements the comet gets into fearful trouble with other heavenly bodies.

In such cases the poor comet is sometimes driven so completely out of its road that it has to make for itself an entirely new path, and our efforts to identify it are plunged in confusion. It has happened that a second comet or even a third will be found in nearly the same track, but whether these are wholly different, or whether they are merely parts of the same original object, it is often impossible to determine.

The great majority of comets are only to be seen with a telescope, and hardly a year passes without the detection of at least a few of these faint objects. The number of really brilliant comets that can be seen in a lifetime could, however, be counted on the fingers.

ENCKE'S COMET.

We have already alluded to a little body called Encke's comet, which was discovered by an astronomer at Marseilles. It was in the year 1818 that he was scanning the heavens with a small telescope, when an object attracted his attention. It was not one of those grand long-tailed comets which every one notices; this body was so faint that it merely appeared as a tiny little cloud of light, and was recognised as a comet by the fact that it was moving about. It happens that there are other bodies in the sky very like comets; we call them nebulæ, and we shall have something to say about them afterwards. But it is remarkable that just as a planet is liable to be mistaken for a star, so a comet is liable to be mistaken for a nebula. However, in each case the fact of its movement is the test by which the planet or the comet is at once detected. A nebula stays always in the same spot, like a star, while a comet is incessantly moving. In fact, with a telescope you can actually watch a comet stealing past the stars that lie near it. You know that an object a very long way off may appear to move slowly, though in reality it is moving very rapidly. Look at a steamer near the horizon at sea. In the course of a minute or two it will not appear to have shifted its position to any appreciable extent, but that is only because it is far off. If you were near the ship you would see that it was dashing along at the rate of perhaps fifteen or twenty miles an hour. In a similar manner the comet seems to move slowly, because it is at such a great distance. In reality, it is moving faster at the time we see it than any steamer, faster than any express train, faster than any cannon-ball. There were special reasons why the movements of Encke's comet should be watched with peculiar care, and the track which it pursued be ascertained. If you can observe a comet three times and measure its position in the sky, the movement of that comet is completely determined. Perhaps I should say would be determined if the comet were let alone, which, unfortunately, is not often the case. Indeed, you may remember how I told you some of the adventures of this very comet when we were speaking about the planet Mercury. Encke's comet comes round in a period of a little more than three years, and it gives us some curious information that has been ascertained during its journeys. One of the facts we have thus learned is so important that we cannot omit to notice it. (Fig. 71.)

At increasing heights above the earth's surface there is gradually less and less air; until at last, at about 200 or 300 miles above the surface on which we dwell, there would be no air. You might as well try to quench your thirst by drinking out of an empty cup as attempt

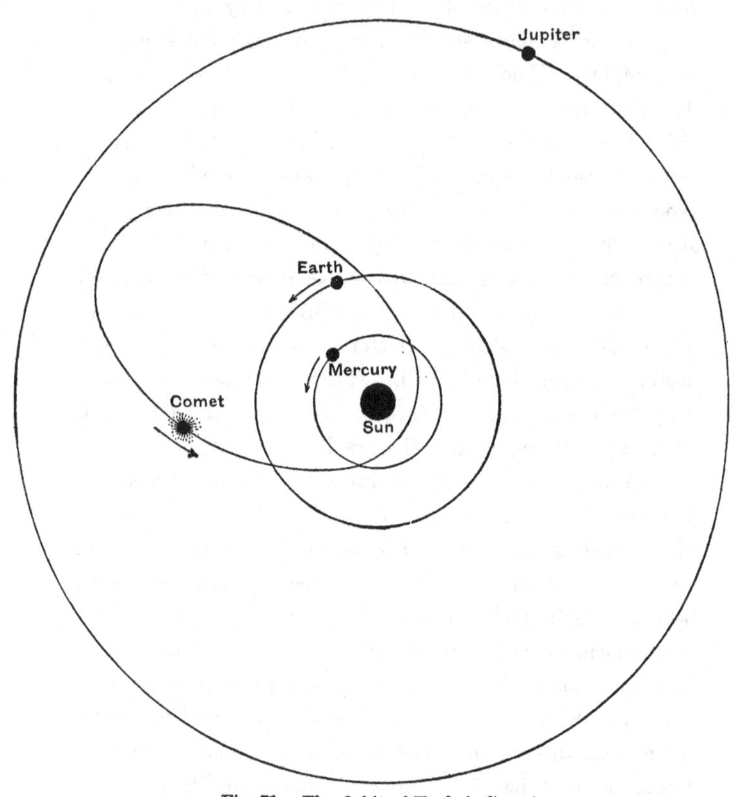

Fig. 71.—The Orbit of Encke's Comet.

to breathe in the open space which begins a few hundred miles aloft. In open space motion could take place quite freely. Down here the resistance of air is a great impediment to movement, especially when very rapid. A heavy cannon-ball is checked and robbed of its pace by having to plough its way through our dense atmosphere. The motion is arrested in the same way, though not of course to the same degree, as if the cannon-ball had been fired into water. Unsubstantial objects are, of course, impeded by the air to a far greater extent than such heavy bodies as cannon-balls are found to be. You know that you cannot throw a handful of feathers across the road in the same way that you could throw a handful of gravel. The light feathers cannot force their way through the air so well as the pebbles. A body so unsubstantial as a comet would never be able to push its way through an atmosphere like ours; but out in empty space the comet meets with no resistance during the greater part of its path. Accordingly, though it has little more substance than a will-o'-th'-wisp, the comet pursues its journey with as much resolute dignity as if it were made of cast iron. If in any part of its track the body should have to pierce its way through any material like even the thinnest possible air, then the unsubstantial nature of cometary materials would be at once shown. The motion would be impeded, and the body's path would be changed. In this way a comet may be made very instructive, for it will show whether space is really so empty as we sometimes suppose it to be. During the greater part of its course the flimsy little Encke tears along with such ease and speed that there can be nothing to impede it, and thus we learn that space is generally empty. However, when the comet begins to wheel around the sun, the freedom of its movements seems to receive a check. The unsubstantial object has to force its way with a difficulty that it did not experience so long as it was moving round the greater part of its orbit. We thus learn that there is a thin diffused atmosphere surrounding the sun. We cannot, indeed, say that it is like our air. Its composition is quite different, and almost the only way we know of its existence is by the evidence which this comet affords. In a former lecture I showed how Encke's comet told us the mass of the planet Mercury. Now we see how the travels of the same body give us information about the sun himself. I ought, however, to add that some more recent comets seem not to have experienced any resistance of the kind we have just been considering.

THE GREAT COMET OF HALLEY.

I dare say you would think it more interesting to talk about some big and bright comets rather than about objects so faint as the comet of Encke. It unfortunately happens that most of the fine comets pay our system only a single visit. There is only one of the really splendid objects of this kind that comes back to us with anything like regularity.

It was last seen in the year 1835, and I am glad to tell you that it is coming again; it is expected about the year 1910. You may ask, How can we feel sure that such a prediction as that I have mentioned will turn out correctly? The fact is that this comet has been watched for a great many centuries. We find ancient records, some of them nearly 2,000 years old, of the appearance of grand comets, and several of these are found to fit in with the supposition that there is a body which accomplishes its journey in a period of about seventy-five or seventy-six years. Of course there are thousands of other comets recorded in these old books as well; but what I mean is that among the records many are found which clearly indicate the successive returns of this particular body.

I will explain how the movements of this comet were discovered. There was a great astronomer called Halley, who lived two centuries ago, and in the year 1682, he, like every one else, was looking with admiration at a splendid comet with a magnificent tail which adorned the sky in that year. At the observatories, of course, they diligently set down the positions of the comet, which they ascertained by carefully measuring it with telescopes. Halley first calculated the highway which this comet followed through the heavens, and then he looked at the list of old comets that had been seen before. He thus found that in 1607—that was, seventy-five years earlier—a great comet had also appeared, the path of which seemed much the same as that which he found for the body that he himself had observed. This was a remarkable fact, and it became still more significant when he discovered that seventy-six years earlier—namely, in 1531—another great comet had been recorded, which moved in a path also agreeing with those of 1607 and 1682. It then occurred to Halley that possibly these were not three different comets, but only different exhibitions of one and the same body, which moved round in the period of seventy-five or seventy-six years.

There is a test which an astronomer can often apply in the proof of his theory, and it is a very severe test. He will not only show himself to be wrong if it fails, but he will also make himself somewhat ridiculous. Halley ventured to submit his reputation to this ordeal. He prophesied that the comet would appear again in another seventy-five or seventy-six years. He knew that he would, of course, be dead long before 1758 should arrive; but when he ventured to make the prediction, he said that he hoped posterity would not refuse to admit that this discovery had been made by an Englishman.

You can easily imagine that as 1758 drew near, great interest was excited among astronomers to see if the prediction of Halley would be fulfilled. We are accustomed in these days to find many astronomical events foretold with the same sort of punctuality as we expect in railway time-tables. The Nautical Almanac is full of such prophecies, and we find them universally fulfilled. Even now, however, we are not able to set forth our time-tables for comets with the same confidence that we show when issuing them for the sun, the moon, or the stars. We do not often find that the Nautical Almanac ventures to be responsible for what freaks comets may take. How astonishing, then, must Halley's prediction have seemed! Here was a vast comet which had to make a voyage through space to the extent of many hundreds of millions of miles. For three-quarters of a century it would be utterly invisible in the greatest telescopes, and the only way in which it could be perceived was by figures and calculations which enabled the mind's eye to follow the hidden body all around its mysterious track. For fifty, or sixty, or seventy years nothing had been seen of the comet, nor, indeed, was anything expected to be seen of it; but as seventy-one, and seventy-two, and seventy-three years had passed, it was felt that the wanderer, though still unseen, must be rapidly drawing near. The problem was made more difficult for those skilful mathematicians who essayed to calculate it by the fact that the comet approached the thoroughfares where the planets circulate; and, of course, the flimsy object would be pulled hither and thither out of its path by the attractions of the weighty bodies. It was computed that the influence of Saturn alone was sufficient to delay the comet for more than three months, while it appeared that the attraction of Jupiter was potent enough to retard the expected event for a year and a half more. Was it not wonderful that mathematicians should be able to find out all these facts from merely knowing the track which the comet was expected to follow? Clairaut, who devoted himself to this problem, suggested that there might also be some disturbances from other causes of which he did not know, and that consequently the expected return of the comet might be a month wrong either way. Great indeed was the admiration of the astronomical world when, true to prediction, the comet faithfully blazed upon the world within the limits of time Clairaut had specified.

The remarkable fulfilment of this prophecy entitles us to speak with confidence about the past performances of this comet. Among all the apparitions of Halley's comet for the last two thousand years, perhaps the most remarkable is that which took place in the year 1066. I am sure you will all remember this date in your English history; it was the year of the Conquest. In those days they did not understand astronomy as we understand it now; they used to think of a comet as a fearful portent of evil, sent to threaten some frightful calamity: such as a pestilence, a war, a famine, or something else equally disagreeable. Hence in the year of the Conquest the appearance of so terrific an object in the sky was a very disagreeable omen. Attention was concentrated upon the spectacle, and a picture of Halley's comet as it appeared to the somewhat terrified imaginations of the people of those days has been preserved. There is a celebrated tapestry at Bayeux on which historical incidents are represented by beautifully worked pictures. On this fabric we have a view of Halley's comet in a quaint and even ludicrous aspect. You will read of this comet also in the early pages of Tennyson's "Harold."

HOW THE TELEGRAPH IS USED FOR COMETS.

In these days the study of comets is prosecuted with energy. Over the world observatories are situated, and whenever a comet is discovered tidings of the event are diffused among those likely to be interested. Suppose that a comet is discovered in the southern hemisphere, the astronomers then write to warn the northern observatories of the event. But comets often move faster than her Majesty's mails, so that the telegraph has to be put into requisition. The kind of message is one which shall show the position and the movements of the comet. It necessarily involves a good many figures and words, and consequently it is desirable to abbreviate as much as possible for the sake of economy. There is a further difficulty in using the telegraph, because the messages are not of an intelligible description to those not specially versed in astronomy. Notwithstanding the well-known skill of the telegraph clerks, they can hardly be expected to be familiar with the technicalities of astronomers. The clerk at the receiving end is handed a message which he does not understand very clearly. The clerk at the other end does not understand the message which is delivered to him, and between them it has happened that they have transformed the message into something which not only they do not understand, but which, unfortunately, nobody else can understand either. All these difficulties have been surmounted by an agreement between astronomers, which is so simple and interesting that I must mention it.

The kind of message that expresses the place of a comet, will consist of sentences something of this kind :—" One hundred and twenty-three degrees and forty-five minutes." Surely it would be an advantage to be able to replace all these words by a single word, particularly if by doing so the risk of error would be diminished. This is what the astronomers' telegraphic arrangement enables them to accomplish. There is a certain excellent Dictionary known as Worcester's. I am sure when Mr. Worcester arranged this work, he had not the slightest expectation of an odd use to which it would occasionally be put. Every astronomer who is co-operating in the comet scheme must have a copy of the book. To send the message I have just referred to, he will take up his Dictionary, and look out page 123. Then he will count down the column until he comes to the forty-fifth word on that page, which he finds to be " constituent," and according to this plan the message, or at least this part of it, is merely that one word, " constituent." The astronomer who receives this message and wishes to interpret it, takes up his copy of Worcester's Dictionary and looks out for " constituent." He sees that it is on page 123, and that it is the forty-fifth word down on that page; and therefore he knows that the interpretation of the message is to be one hundred and twenty-three degrees and forty-five minutes.

THE PARABOLA.

Generally speaking, great comets come to us once and are then never seen again. Such bodies do not move in closed ovals or ellipses, they follow another kind of curve, which is represented in Fig. 57. It is one that every boy ought to know. In fact, in one of his earliest accomplishments, he learned how to make a parabola. It is true he did not call it by any name so fine as this, but

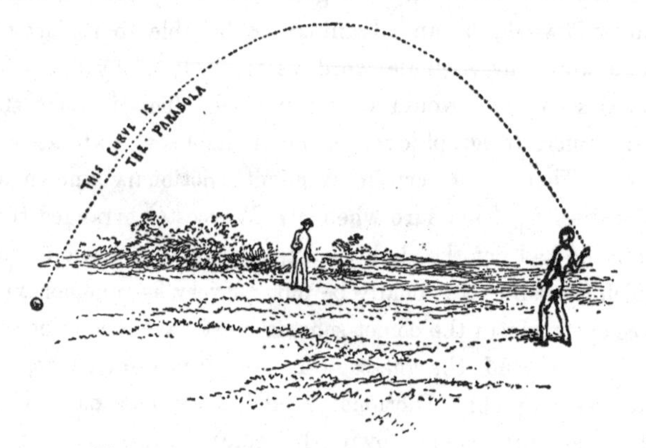

Fig. 72.—The Path of a Projectile is a Parabola.

every time a ball is thrown into the air it describes a part of that beautiful curve which geometers know by this word (Fig. 72). In fact, you could not throw a ball so that it should describe any other curve except a parabola. No boy could throw a stone in a truly horizontal line. It will always curve down a little, will always, in fact, be a portion of a parabola.

There are big parabolas and there are small ones. One of the shells which are thrown into a town when bombarded from a distance describes, as it rises and then slopes down again, part of a mighty parabola. So does a tennis ball thrown by the hand or struck by the racket; though here, indeed, I admit that a spin may be given to the ball which will somewhat detract from the simplicity of its movement. In playing base-ball, a large part of the skill of the pitcher consists in throwing the ball in such a way that it shall not move in a parabola, but in some twisting curve by which he hopes to baffle his adversary. Setting aside these exceptions, and such another exception as the case of a body tossed straight up or dropped straight down, we may assert that the path of a projectile is a parabola.

There are some remarkable applications of the same curve for practical purposes. From our lighthouses we want to send beams off to sea, so as to guide ships into port. If we merely employed a lamp without concentrating its rays, we should have a very imperfect lighthouse, for the lamp scatters light about in all directions. Much of it goes straight up into the air, much of it would be directed inland; in fact, there is only an extremely small part of the entire number of rays that will naturally go in the useful direction. We therefore require something round the lamp which shall catch the truant rays that are running away to idleness and loss, and shall concentrate them into the direction in which they will be useful to the mariner. One of the ways of doing this is to furnish the lamp with a reflector. On its bright surface (Fig. 73) all the rays fall which would otherwise have gone astray, and each of them is simply redirected, where the sailors can see it. It is essential that the mirror shall do this work accurately, and this it will only do when it has been truly shaped so as to be a parabola.

You will remember, also, how I described to you the reflector which Herschel made for his great telescope (Fig. 67). The shape of the mirror must be most accurately worked, and it, too, must have a parabola for its section; so that you see this curve is one of importance in a variety of ways.

But the grandest of all parabolas are those which the

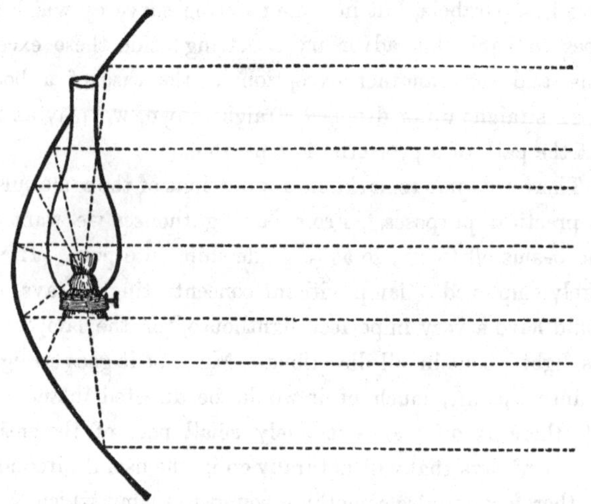

Fig. 73.—The Lighthouse Reflector.

comets pursue. Unlike the ellipse, the parabola is an open curve; it has two branches stretching away and away for ever, and always getting further apart. Of course, in the examples of this curve that I have given, it is only a small part of the figure that is concerned. Where you throw a stone it only describes that part of the parabola that lies between your hand and the spot where the stone hits the ground. It is just a part of the curve in the same way that a crescent may be a bit of a circle. It is to comets we must look for the most complete illustration of the ample extent of a parabola.

The shape of this grand curve will explain why so many comets only appear to us once. It is quite clear that if you begin to run round a closed race-course, you may, if you continue your career long enough, pass and repass the starting post thousands of times. Thus, comets which move in ellipses, and are consequently tracing closed curves, will pass the earth times without number. Thus we may see them over and over again, as we do Encke's comet or Halley's comet. But suppose you were travelling along a road which, no matter how it may turn, never leads again into itself, then it is quite plain that, even if you were to continue your journey for ever, you can never twice pass the same house on the roadside. That is exactly the condition in which most of the comets are moving. Their orbits are parabolas which bend round the sun; and, generally speaking, the sun is very close to the turning-point. The earth is also, comparatively speaking, close to the sun; so that while the comet is in that neighbourhood we can sometimes see it. We do not see the comet for a long time before it approaches the sun, or for a long time after it has passed the sun. All we know, therefore, of its journey is that the two ends of the parabola stretch on and on for ever into space. The comet is first perceived coming in along one of these branches to whirl round the sun; and after doing so, it retreats again along the other branch, and gradually sinks into the depths of space.

Why one of these mysterious wanderers should approach in such a hurry, and then why it should fly back again, can be partially explained without the aid of mathematics.

Let us suppose that, at a distance of thousands of millions of miles, there floated a mass of flimsy material resembling that from which comets are made. Notwithstanding its vast distance from the sun, the attraction of that great body will extend thither. It is true the pull of the sun on the comet will be of the feeblest and slightest description, on account of the enormously great distance. Still, the comet will respond in some degree, and will commence gradually to move in the direction in which the sun invites it. Perhaps centuries, or perhaps thousands, or even tens of thousands, of years will elapse before the object has gained the solar system. By that time its speed will have been augmented to such a degree, that after a terrific whirl around the sun, it will at once fly off again along the other branch of the parabola. Perhaps you will wonder why it does not tumble straight into the sun. It would do so, no doubt, if it started at first from a position of rest; generally, however, the comet has a motion to begin with which would not be directed exactly to the sun. This it is which causes the comet to miss actually hitting the sun.

It may also be difficult to understand why the sun does not keep the comet when at last it has arrived. Why should the comet be in such a hurry to recede? Surely it might be expected that the attraction of the sun ought to hold it. If something were to check the pace of the comet in its terrific dash round the sun, then, no doubt, the comet would simply tumble down into the sun and be lost. The sun has, however, not time to pull in the comet when it comes up with a speed 20,000 times that of an express train. But the sun does succeed in altering the *direction* of the motion of the comet, and the attraction has shown itself in that way.

I can illustrate what happens in this manner. Here is a heavy weight suspended from the ceiling by a wire; it hangs straight down, of course, and there it is kept by the pull of the earth. Supposing I draw the weight aside and allow it to swing to and fro, then the motion continues like the beat of a pendulum. The weight is always pulled down as near to the earth as possible, but when it gets to the lowest point, it does not stay there, it goes through that point and rises up at the other side. The reason is that the weight has acquired speed by the time it reaches the lowest point; and that, in virtue of its speed, it passes through the position in which it would naturally rest, and actually ascends the other side in opposition to the earth's pull, which is dragging it back all the time. This will illustrate how the comet can pass by and even recede from the body which is continually attracting it.

Just a few words of caution must be added. Suppose you had an ellipse so long that the comet would take thousands and thousands of years to complete a circuit, then the part of the ellipse in which the comet moves during the time when we can see it is so like a parabola, that we might possibly be mistaken in the matter. In fact, a geometer will tell us that if one end of an ellipse was to go further and further away, the end that stayed with us would gradually become more and more like a parabola. Therefore, some of those comets which seem to move in parabolas, may really be moving in extremely elongated ellipses; and, thus, after excessively long periods of time may come back to revisit us.

THE MATERIALS OF A COMET.

A comet is made of very unsubstantial material. This we can show in a very interesting manner, when we see the comet moving over the sky between the earth and the stars. Sometimes a comet will pass over a cluster of very small stars, so faint that the very lightest cloud that is ever in the sky would be quite sufficient to hide them. Yet the stars are distinctly visible right through the comet, notwithstanding that it is hundreds of thousands of miles thick. This shows us how excessively flimsy is the substance of a comet, for while a few feet of haze or mist suffices to extinguish the brightest of stars, this immense curtain of comet stuff, whatever it may be made of, is practically transparent.

I have often told you that we are able to weigh the heavenly bodies, but a comet gives us a great deal of trouble. You see that the weighing machine must be of a very delicate kind if you are going to weigh a very light object. Take, for example, a little lock of golden hair, which no doubt has generally a value quite independent of the number of grains that it contains. Suppose, however, that we are so curious as to desire to know its weight, then one of those beautiful balances in our laboratories will tell us. In fact, if you snipped a little fragment from a single hair, the balance would be sensitive enough to weigh it. If, however, you were only provided with a common pair of scales like those which are suited for the parcel post, then you could never weigh anything so light as a lock of hair. You have not small enough weights to begin with, and even if you had they would be of no use, for the scale is too coarse to estimate such a trifle. This is precisely the sort of difficulty we experience when we try to weigh a comet. The body, though so big, is very light, and our scales are so coarse that we are in a position of one who would try to weigh a lock of hair with a parcel-post balance. We cannot always find a suitable pair of scales for weighing celestial bodies. We have to use for the purpose whatever methods of discovering the weights happen to be available. So far the methods I have mentioned are of the coarsest description, they serve well enough for weighing heavy masses like planets, but they will not do for such unsubstantial bodies as comets.

But, though we fail in this endeavour, yet skilful astronomers have succeeded in something which at first you might think to be almost impossible. They have actually been able to discover some of the ingredients of which a comet is made. This is so important a subject that I must explain it fully.

The most instructive comet which we have seen in modern days is that which appeared in the year 1882. It was an object so great that its tail alone was double as long as from the earth to the sun. It was discovered at the observatories in the southern hemisphere early in the September of that year. A little later it was observed in the northern hemisphere under extraordinary circumstances. It must be remembered that a comet is generally a faint object, and that even those comets which are large enough and bright enough to form glorious spectacles in the sky when dark, are usually invisible during the brightness of day. For a comet to be seen in daylight was indeed an unusual occurrence; but on the forenoon of Sunday, September 17th, Mr. Common at Ealing saw a great comet close to the sun. Unfortunately clouds intervened, and he was prevented from observing the critical occurrence just approaching. An astronomer at the Cape of Good Hope—Mr. Finlay—who had also been one of the earliest discoverers of the comet, was watching the body on the same day. He followed it as it advanced close up to the sun: bright indeed must that comet have been which permitted such a wonderful observation. At the sun's edge the comet disappeared instantly; in fact, the observers thought that it must have gone behind the sun. They could not otherwise account for the suddenness with which it vanished. This was not what really happened. It was afterwards ascertained that the comet had not passed behind the sun; it had indeed come between us and the sun. In its further progress this body showed in a striking degree the incoherent nature of the materials of which a comet is composed. It seemed to throw off portions of its mass along its track, each of which continued an independent journey. Even the central part in the head of the comet—the nucleus, as it is called—showed itself to be of a widely different nature from a substantial planetary body. The nucleus divided into two, three, four, or even five distinct parts, which seemed, in the words of one observer, to be connected together like pearls on a string.

The comet of 1882 was also very instructive with regard to the actual materials from which such bodies are made. Astronomers have a beautiful method by which they find out the substances present in a heavenly body, even though they never can get a specimen of the body into their hands. We know at least three substances which were present in this comet. The first of them is an ingredient which is very commonly found in comets—a chemist calls it carbon. It is an extremely familiar material on the earth ; for instance, coal is chiefly composed of carbon. Charcoal when burned leaves only a few ashes. All the substance that has vanished during combustion is carbon ; in fact, it is not too much to say that carbon is found abundantly not only in wood, but in almost every form of vegetable matter. The food we eat contains abundant carbon, and it is an important constituent in the building up of our own bodies. Generally speaking, carbon is not found in a pure state—it is associated with other substances. Soot and lampblack are largely composed of it ; but the purest form of this element that we know is the diamond.

It is interesting to note that carbon is certainly found as a frequent constituent of comets. The great comet of 1882 undoubtedly contained it, as well as certain other substances. Of these we know two : the first is the element sodium, an extremely abundant material on earth, inasmuch as the salt in the sea is mainly composed of it. It was also discovered that the same great comet contained another substance extremely abundant here and extremely useful to mankind. Dr. Copeland and Dr. Lohse at Dunecht showed that iron was present in this body which had come in to visit us from the depths of space.

These discoveries are specially interesting, because they seem to show the unity of material in our system. We already knew that sodium and that iron abounded in the sun, and now we have learned that these bodies and carbon as well are present in the comets. In the next chapter we shall learn that the very same materials—sodium and iron—are met with in bodies far more remote from us than any bodies of our own system.

Comets have such a capricious habit of dashing into the solar system at any time and from any direction, that it is worth while asking whether a comet might not sometimes happen to come into collision with the earth. There is nothing impossible in this occurrence. There is, however, no reason to apprehend that any disastrous consequences would ensue to the earth. Man has lived on this globe for many, many thousands of years, and the rocks are full of the remains of fossil animals which have flourished during past ages; indeed, we cannot attempt to estimate the number of millions of years that have elapsed since living things first crawled about this globe. There has never been any complete break in the succession of life, consequently during all those millions of years we are certain that no such dire calamity has happened to the earth as a frightful collision would have produced, and we need not apprehend any such catastrophe in the future.

I do not mean, however, that harmless collisions with comets may not have occasionally happened; in fact, there is good reason for knowing that they have actually taken place. In the year 1861 a fine comet appeared; it is not so well remembered as its merits deserve, because it happened, unfortunately for its own renown, to appear just three years after the comet of 1858, which was one of the most gorgeous objects of this kind in modern times. But in 1861 we had a novel experience. On a Sunday evening in midsummer of that year, we dashed into the comet, or it dashed into us. We were not, it is true, in collision with its densest part; it was rather the end of the tail which we encountered. There were, fortunately, no very disastrous consequences. Indeed, most of us never knew that anything had happened at all, and the rest only learned of the accident long after it was all over. For a couple of hours that night it would seem that we were actually in the tail of the comet, but so far as I know no one was injured or experienced any alarming inconvenience. Indeed, I have only heard of one disaster arising from the collision. A clergyman tells us that at midsummer he was always able in ordinary years to read his sermon at evening service without artificial light. On this particular occasion, however, the sky was overcast with a peculiar glow, while the ordinary light was so much interfered with that the sexton had to provide a pair of candles to enable him to get through the sermon. The expense of those candles was, I believe, the only loss to the earth in consequence of its collision with the comet of 1861.

The tail of a comet appears to develop under the influence of the sun. As the wandering body approaches the source of central heat it grows warm, and as it gets closer and closer to the sun, the fervour gets greater and greater, until sometimes the comet experiences a heat more violent than any we could produce in our furnaces. The most infusible substances, such as stones or earth, would be heated white-hot and melted, and even driven off into vapour, under the intense heat to which a comet is sometimes exposed. Comets, indeed, have been known to sweep round the sun so closely as to pass within a seventh part of its radius from the surface. It seems that certain materials present in the comet, when heated to this extraordinary temperature, are driven away from the head, and thus form the tail (Fig. 74). Hence we see that the tail consists of a stream of vaporous particles, dashing away from the sun as if the heat which had called them

Fig. 74.—How the Tail of a Comet arises.

into being was a torment from which they were endeavouring to escape.

The tail of a comet is, therefore, not a permanent part of the body. It is more like the smoke from a great chimney. The smoke is being incessantly renewed at one end as the column gets dispersed into the air at the other. As the comet retreats, the sun's heat loses its power. In the chills of space there is, therefore, no tail-making in progress, while the small mass of the comet renders it unable to gather back again by its attraction the materials which have been expelled. Should it happen that the comet moves in an elliptic orbit, and thus comes back time after time to be invigorated by a good roasting from the sun, it will of course endeavour to manufacture a tail each time that it approaches the source of heat. The quantity of material available for the formation of tails is limited to the amount with which the comet originally started; no fresh supply can be added. If, therefore, the comet expends a portion of this every time it comes round, an inevitable consequence seems to follow. Suppose a boy receives a sovereign when he goes back to school, and that every time he passes the pastrycook's shop some of his money disappears in a manner that I dare say you can conjecture, I need not tell you that before long the sovereign will have totally vanished. In a similar way comets cannot escape the natural consequences of their extravagance; their store of tail-making substance must, therefore, gradually diminish. At each successive return the tails produced must generally decline in size and magnificence, until at last the necessary materials have been all squandered, and we have the pitiful spectacle of a comet without any tail at all.

The gigantic size of comets must excite our astonishment. A pebble tossed into a river would not be more completely engulfed than is our whole earth when it enters the tail of one of these bodies. But we now pass by a sudden transition to speak of the very smallest bodies, of little objects so minute that you could carry them in your waistcoat pocket. You will perhaps be surprised that such things can play an important part in our system, and have an important connection with mighty comets.

METEORS.

If you look out from your window at the midnight sky, or take a walk on a fine clear night, you will occasionally see a streak of light dash over the heavens, thus forming what is called a falling star or a shooting star (Fig. 75). It is not really one of the regular stars that has darted from its place. The objects we are now talking of are quite different from

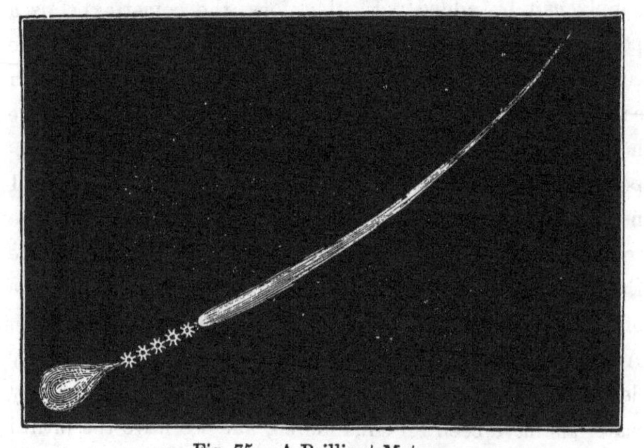

Fig. 75.—A Brilliant Meteor.

stars proper. To begin with, the shooting stars are comparatively close to us, and they are very small, whereas the stars themselves are enormous globes, far bigger than our earth, or often even bigger than the sun. Sometimes a great shooting star is seen which makes a tremendous blaze of light as bright as the moon, or even brighter still. These objects we call meteors, and you will be very fortunate if you can ever see a really fine one. Astronomers cannot predict these things as they predict the appearance of the planets. Bright meteors consequently appear quite unexpectedly, and it is a matter of chance as to who shall enjoy the privilege of beholding them. But it is not about the great meteors that we are now going to speak particularly; they are often not so interesting as the small ones.

These little meteoroids, as we shall call them, have a curious history. They become visible to us only at the very last moment of their existence—in fact, the streak of light which forms a shooting star is merely the destruction of a meteoroid. You must always remember that we are here living at the bottom of a great ocean of air, and above the air extends the empty space. Air is a great impediment to motion; a large part of the power of a locomotive engine has to be expended solely in pushing the air out of the way so as to allow the train to get through. The faster the speed, the greater is the tax which the air imposes on the moving body. A cannon-ball, for instance, loses an immensity of its speed, and consequently of its power, by having to bore its way through the air. In outer space beyond the limits of this atmosphere, a freedom of movement can be enjoyed of which we know nothing down here. I spoke of this when discussing the movements of Encke's comet. Even this very unsubstantial body could dash along without appreciable resistance until it traversed the atmosphere surrounding the sun. But now we have to speak of the motion of a little object both small and dense, resembling perhaps a pebble or a small bit of iron, or some substance of that description. It is a little body of this kind which produces a shooting star.

For ages and ages the meteoroid has been moving freely

through space. The speed with which it dashes along greatly exceeds that of any of the motions with which we are familiar. It is about 100 times as swift as the pace of a rifle-bullet. About twenty miles would be covered in a second. You can hardly imagine what that speed is capable of. Suppose that you put one of these flying meteoroids beside an express train to race from London to Edinburgh, the meteoroid would have won the race before the train could have got out of the station. Or suppose that a shooting star determined to make the circuit of the earth, it might, so far as pace is concerned, go entirely around the globe and back to the point from which it started in a little more than twenty minutes. But the fact is, you could not make any object down here move as fast as a shooting star. No gunpowder that could be made would be strong enough, to begin with; and even if the body could once receive the speed, it would never be able to force its way through the air uninjured.

So long as a little shooting star is tearing away through open space we are not able to see it. The largest telescope in the world would not reveal a glimpse of anything so small. The meteoroid has no light of its own, and it is not big enough to exhibit the light reflected from the sun in the same manner as a little planet would do. It is only at the moment when it begins to be destroyed that its visibility commences. If the little object can succeed in dashing past our earth without becoming entangled in the atmosphere, then it will pursue its track with perhaps only a slight alteration in its path, due to the pull exercised by the earth. The atmosphere which surrounds our globe may be likened to a vast net, in which if any little meteor becomes caught its career is over. For when the little body, after rejoicing in the freedom of open space, dashes into air, immediately it experiences a terrific resistance ; it has to force the particles of air out of the way so as to make room for itself, and in doing so it rubs against them with such vehemence that heat is produced.

I am sure every boy knows that if he rubs a button upon a board it becomes very hot, in consequence of the friction. There are many other ways in which we can illustrate the production of heat in the same manner. One is a contrivance by which we spin round rapidly a piece of stick pressed against a board. Quantities of heat are thus produced by the friction, and volumes of smoke rise up. We have read how some savages are able to produce fire by means of friction in a somewhat similar manner, but to do so requires a rare amount of skill and patience. There is another illustration by which to show how heat can be produced by friction. A brass tube full of water is so arranged that it can be turned around very rapidly by the whirling table. We apply pressure to the tube, and after a minute or two the water begins to get hot, and then at last to boil, until ultimately the cork is driven out and a diminutive and, fortunately, harmless explosion of the friction boiler takes place. Engineers are aware how frequently heat is produced by friction, when it is very inconvenient or dangerous. Indeed, unless the wheels of railway carriages are kept well greased, the rubbing of the axle may generate so much heat that accidents will ensue. Nature, in the little shooting star, gives us a striking illustration of the same fact. Perhaps you may be surprised to hear that the whole brilliancy of the shooting star is simply due to friction. As the little body dashes through the air it becomes first red-hot, then white-hot, until at last it is melted and turned into vapour. Thus is formed that glowing streak which we, standing very many miles below, see as a shooting star.

A bullet when fired from a rifle will fly into pieces after it has struck against the target, and if you pick up one of these pieces soon after the shot you will generally find it quite hot. Whence comes this heat? The bullet, of course, was cold before the rifleman pulled the trigger. No doubt there was a considerable amount of heat developed by the burning of the gunpowder, but the bullet was so short a time in contact with the wad, through which so little heat would pass, that we must look to some other source for the warmth that has been acquired. Friction against the barrel as the bullet passed to the mouth must have warmed the missile a good deal, and when rubbing against the air the same influence must have added still further to its temperature, while the blow against the target would finally warm it yet more.

In comparing the shooting star with the rifle-bullet we must remember that the celestial object is travelling with a pace 100 times as swift as the utmost velocity that the rifle can produce, and the heat which is generated by friction is increased in still greater proportion. If we double the speed we shall increase the quantity of heat by friction fourfold; if we increase the speed three times then friction will be capable of producing nine times as much heat. In fact, we must multiply the number expressing the relative speed by itself—that is, we must form its square—if we want to find an accurate measure for the quantity of heat which friction is able to produce when a rapidly moving body is being stopped by its aid. The shooting star may have a pace 100 times that of the rifle-bullet, and if we multiply 100 by 100 we get 10,000; consequently we see that the heat produced by the shooting star before its motion was arrested by dashing through the air would be 10,000 times that gained by the rifle-bullet in its flight. If the temperature of the rifle-bullet only rose a single degree by friction, it would thus be possible for the shooting star to gain 10,000 degrees, and this would be enough to melt and boil away any object which ever existed. Thus we need not be surprised that friction through the air, and friction alone, has proved an adequate cause for the production of all the heat necessary to account for the most brilliant of meteors.

It is rather fortunate for us that the meteors do dash in with this frightful speed; had the little bodies only moved as quickly as a rifle-bullet, or even only four or five times as fast, they would have pelted down on the earth in solid form. Indeed, on rare occasions it does happen that bodies from the heavens strike down on the ground. The great majority of these little bodies, however, become entirely transformed into harmless vapour. The earth would, indeed, be almost uninhabitable from this cause alone were it not for the protection that the air affords us. All day and all night innumerable missiles would be whizzing about us, and though many of them are undoubtedly very small, yet as their speed is 100 times that of a rifle bullet, the fusil-lade would be very unpleasant. It is, indeed, the intense hurry of these celestial bullets to get at us which is the very source of our safety. It dissipates the meteors into streaks of harmless vapour.

WHAT BECOMES OF THE SHOOTING STARS.

When we throw a lump of coal on the fire, all that is soon left is a little pinch of ashes, and the rest of the coal has vanished. You might think it had been altogether annihilated, but that is not nature's way. Nothing is ever completely destroyed; it is merely transformed or changed into something else. The greater part of the coal has united with the oxygen which it has obtained from the air, and has formed a new gas, which has ascended the chimney. Every particle that was in the coal can be thus accounted for, and in the act of transformation heat is given out.

A meteor also becomes transformed, but the substance it contains is not lost, though it is changed into glowing vapours. It is known that with heat enough any substance can be turned into vapour, just as water can be boiled into steam. Look at an electric light flashing between two pieces of carbon. Though carbon is one of the most difficult substances to melt, yet such is the intense heat generated by the electric current that the carbon is not only melted, but is actually turned into a vapour, and it is this vapour glowing with heat that gives us the brilliant light. In a similar manner iron can be rendered red-hot, white-hot, and then boiled and transformed into an iron vapour, if we may so call it. There is, indeed, plenty of such iron vapour in the universe. Quantities of it surround the sun and some of the stars.

When ordinary steam is chilled it condenses into little drops of water. So too, if iron be heated until it is transformed into vapour, and if that vapour be allowed to condense, it will ultimately form a dust, consisting of bits of iron so small that you would require a microscope to examine them. There is iron present in the small shooting stars. Other substances are also contained therein, and all these materials, after being boiled up by the intense heat, are transformed into vapour. When the heat subsides, the vapour condenses again into fine dust, so that the ultimate effect of the atmosphere on a shooting star is to grind the little object into excessively fine powder or dust, which is scattered along the track which the object has pursued. Sometimes this dust will continue to glow for minutes after the meteor has vanished, and in the case of some great meteors this stream of luminous dust in the air forms a very striking spectacle. A great meteor, or fire-ball as it is often called, appeared on the 6th of November, 1869. It flew over Devonshire and Cornwall, and left a track fifty miles long and four miles wide. The dust remained visible all along the great highway for nearly an hour; it formed a gloomy cloud hanging in the sky, and though originally nearly straight, it became bent and twisted by the winds before it finally disappeared from view.

We have now to see what becomes of this meteoric dust which is being incessantly poured into the air from external space. None of it ever gets away again; for whenever an unfortunate meteor just touches the air it is inevitably captured and pulverised. That dust subsides slowly, but we do not find it easy to distinguish the particles which have come from the shooting stars, because there is so much floating dust which has come from other sources.

A sunbeam is the prettiest way of revealing the existence of the motes with which the air is charged. The sunbeam renders these motes visible exactly in the same way as planets become visible when sunbeams fall on them in far-distant space. But if we have not the sunbeams here, we can throw across the room a beam of electric light, and it is seen glowing all along its track, simply because the air of the room, like air everywhere, is charged with myriads of small floating particles. If you hold the flame of a spirit-lamp beneath this beam, you will see what seems like columns of black smoke ascending through it. But these columns are not smoke, they are pure air, or rather air in which the solid particles have been transformed into vapour by the heat from the spirit-flame.

The motes abound everywhere in the air. We take thousands of them into our lungs every time we breathe. They are on the whole gradually sinking and subsiding downwards, but they yield to every slightest current, so that when looking at a sunbeam you will find them moving in all directions. It is sometimes hard to believe that the little objects are tending downwards, but if you look on the top of a book that has lain for a time on a book-shelf you find there a quantity of dust, produced by the motes which have gradually subsided where they found a quiet spot and were allowed sufficient time to do so.

The great majority of these particles consist, no doubt, of fragments of terrestrial objects. The dust from the roads, the smoke from the factories, and numerous other sources, are incessantly adding their objectionable particles to the air. There can be no doubt that the shooting stars also contribute their mites to the dust with which the atmosphere is ever charged. The motes in the murky atmosphere of our towns have no doubt chiefly originated from sources on this earth. Many of these sources it would be impossible to regard as of a romantic description.

We may, however, feel confident that among those teeming myriads are many little particles which, having had their origin from shooting stars, are now gradually sinking to the earth.

This is not a mere surmise, for dust has been collected from lofty Alpine snows, from the depths of the sea, and from other localities far removed from the haunts of men. From such collections, tiny particles of iron have been obtained, which have evidently been once in a molten condition. There is no conceivable explanation for the origin of iron fragments in such situations, except that they have been dropped from shooting stars.

I am sure you have often helped in the making of a gigantic snowball. You begin with a small quantity of snow that can be worked with your hands. Then you have rolled it along the ground until it has become so big and so heavy, that you must get a few playmates to help you, until at last it has grown so unwieldy that you can move it no longer, and then you apply your artistic powers to carving out a statue. The snowball has grown by the addition of material to it from without, and as it became heavier and heavier, it lapped up more and more of the snow as it rolled along; so that with each increase of size, its capacity for becoming still larger has also increased. I want to liken our earth to a snowball, which goes rolling on through space, and every day, every hour, every minute, is gathering up and taking into it the little shooting stars that it meets with on its way. No doubt the annual gathering is a very small quantity when compared with the whole size of the earth; but the important point is that the earth is always drawing in, and now, at all events, never giving back again; so that when this process goes on long enough, astonishing results may be obtained.

You have all heard many maxims on this subject—how every little saving will at length reach a respectable or a gigantic total. Nature abounds with illustrations of the principle. All the water that thunders over Niagara is merely a sufficient number of little drops of rain collected together. Our earth has been gradually hoarding up, during countless ages, all the meteor dust that has rained upon it; and the larger the earth grows, the bigger is the net which it spreads, and the greater is the power it has to capture the wandering meteors. Thus, our earth, ages and ages ago, may have been considerably smaller than it is at present; in fact, a large proportion of this globe may have been derived from the little shooting stars which incessantly rain in upon its surface.

GRAND METEORS.

I dare say that many of those present will, in the course of their lives, have opportunities of seeing some of the great meteors, or fire-balls, which are occasionally displayed. Generally speaking, about one hundred or so of these splendid objects are recorded every year. We are never apprised that they are coming; they take us unawares, and therefore we have no opportunity to make proper arrangements for seeing them. It is only a chance that such persons as have been fortunate enough to see them, will have noted the circumstances with sufficient accuracy to enable us to make use of their observations.

The chief point to determine is the height of the meteor above the earth. For this we must have two observations, at least, made in places as far asunder as possible. Suppose an observer at London and an observer at York were both witnesses of a splendid meteor; if they find, on subsequent comparison, that their observations were made at the same moment, there is no reasonable doubt that it was the same object they both saw. The observer at York describes the meteor as lying to the south, and that it was half-way down

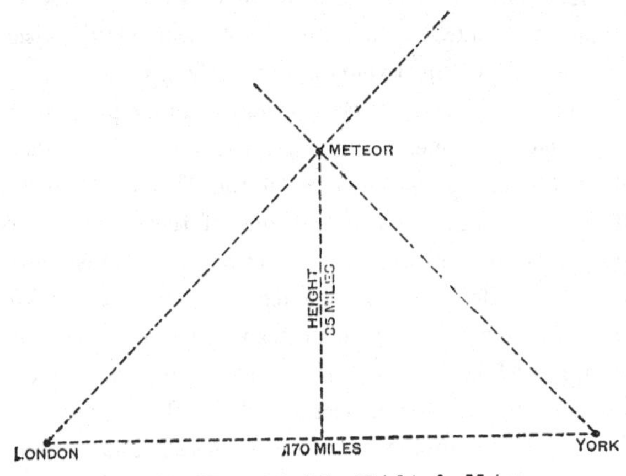

Fig. 76.—How to find the Height of a Meteor.

from the point directly over his head towards the horizon. The London observer speaks of the meteor as being to the north; and to him also it appeared that the object was half-way down towards his horizon from the point directly over his head. If you know a little Euclid you can easily show from these facts that the height of the meteor must have been half the distance between London and York, that is 85 miles (Fig. 76).

I do not mean to say that the mode of discovering the

meteor's height will be always quite such a simple process as it has been in the case of the London and York observations. The principle is, however, the same—that whenever from two sufficiently distant positions the direction of the meteor has been observed its path is known, just as in p. 20 we showed how the height of the suspended ball was obtained from observations at each end of the table. Generally speaking, bright meteors begin at an elevation of between fifty and one hundred miles, and they become extinguished before they are within twenty miles of the ground.

Sometimes a tremendous explosion will take place during the passage of a meteor through the air. There was a celebrated instance in America on the 21st of December, 1876, which will give an idea of one of these objects possessing exceptional magnificence. It began in Kansas about seventy-five miles high, and thence it flew for a thousand miles at a speed of ten or fifteen miles a second, until it disappeared somewhere near Lake Ontario. Over a certain region between Chicago and St. Louis, the great ball of fire burst into a number of pieces, and formed a cluster of glowing stars that seemed to chase each other over the sky. This cluster must have been about forty miles long and five miles wide, and when the explosion occurred a most terrific noise was produced, so loud that many thought it was an earthquake. A remarkable circumstance illustrates forcibly the tremendous height at which this explosion occurred. The meteor had burst into pieces, the display was all over and was beginning to be forgotten, and yet nothing had been *heard*. It was not until a quarter of an hour after the explosion had been *seen* that a fearful crash was heard at Bloomington. It was 180 miles from the spot where the explosion actually occurred, and as sound takes five seconds to travel a mile, you can easily calculate that the mighty noise required a quarter of an hour for its journey.

Shooting stars are of every grade of brightness. Beginning with the more gorgeous objects which have been compared with the moon or even with the sun himself, we descend to others as bright as Venus or as Jupiter; others are as bright as stars of various degrees of brilliancy. Fainter shooting stars are much more numerous than the conspicuous ones; in fact, there are multitudes of these objects so extremely feeble that the unaided eye would not show them. They only become perceptible in a telescope. It is not uncommon while watching the heavens at night to notice a faint streak of light dashing across the field of the instrument. This is a shooting star, which is invisible except through the telescope.

THE GREAT NOVEMBER SHOWERS.

Occasionally we have the superb spectacle of a shower of shooting stars. None of you, my young friends, can as yet have had the good fortune to witness one of the specially grand displays, but you may live in hope; there are still showers to come. Astronomers have ventured on the prophecy that in or about the year 1899 you will have the opportunity of seeing a magnificent exhibition of this kind. There is only one ground for anxiety, and that is as to whether the clouds will keep out of the way for the occasion. I think I cannot explain my subject better than by taking you into my confidence and showing you the reasons on which we base this prediction. The last great shooting-star shower took place in the year

1866, or, perhaps, I should rather say that this was the last display from the same shooting star system as that about which we are now going to speak. On the night of the

Fig. 77.—A Great Shower of Shooting Stars.

13th November, 1866, astronomers were everywhere delighted by a superb spectacle. Enjoyment of the wondrous sight was not only for astronomers. Everyone who loves to see the great sights of nature will have good reason for remembering that night. I certainly shall never forget it. It was about ten o'clock when a brilliant meteor or two first flashed across the sky, then presently they came in twos and threes, and later on in dozens, in scores, in hundreds. These meteors were brilliant objects, any one of which would have extorted admiration on an ordinary night. What then was the splendour of the display when they came on in multitudes? For two or three hours the great shower lasted, and then gradually subsided.

We were not taken unawares on this occasion, for the shower was expected, and had been, in fact, awaited with eager anticipation. It should first be noticed that on the recurrence each year of the 13th of November some shooting stars may always be looked for. Every thirty-three years, or thereabouts, the ordinary spectacle breaks out into a magnificent display. It has also been found that for nearly 1,000 years there have been occasional grand showers of meteors at the time of year mentioned, and all these incidents agree with the supposition that they are merely repetitions of the regular thirty-three-year shower. The first was in the year A.D. 902, which an old chronicle speaks of as the "year of the stars," from the extraordinary display which then took place. I do not think the good people 1,000 years ago fully appreciated the astronomical interest of such spectacles; in fact, they were often frightened out of their wits, and thought the end of the world had come. Doubtless many ancient showers have taken place of which we have no record whatever. In more modern days we have somewhat fuller information; for example, on the night between the 12th and 13th of November, 1833, a shower was magnificently seen in America. Mr. Kirkwood tells us that a gentleman of South

Carolina described the effect on the negroes of his plantation as follows:—"I was suddenly awakened by the most distressing cries that ever fell on my ears. Shrieks of horror and cries for mercy I could hear from most of the negroes of the three plantations, amounting in all to about 600 or 800. While earnestly listening for the cause, I heard a faint voice near the door calling my name. I arose, and taking my sword, stood at the door. At this moment I heard the same voice still beseeching me to arise, and crying out that the world was on fire. I then opened the door, and it is difficult to say which excited me the most—the awfulness of the scene or the distressed cries of the negroes. Upwards of a hundred lay prostrate on the ground, some speechless, and some with the bitterest cries, but with their hands raised praying for mercy. The scene was truly awful, for never did rain fall much thicker than the meteors fell towards the earth."

By the study of many records of great showers it was learned that the intervals at which these grand displays succeeded one another was about thirty-three years; and when it was remembered that the last great shower was in 1833 it was confidently expected that another similar display would take place in 1866. This was splendidly confirmed. Yet another thirty-three years will bring us to 1899, when we have good reason for expecting that a grand shower of these bodies may be once again expected on November 13th of that year. It may, however, possibly be that the shower will occur on the same day of the succeeding year, or possibly on both occasions.

We know a good deal now as regards the movements of these little objects. I want you to think of a vast swarm, something like a flock of birds, which I dare say you have often seen flying high in the air; the difference, however, is that the flock of meteors is enormously greater than any flock of birds ever was; and the meteors, too, are scattered so widely apart, that each one may be miles away from its next neighbours. Usually the meteoric shoal is many millions of miles long, and perhaps a hundred thousand miles in width. The great flock of meteors travels through space in a certain definite track. We have learned how the sun guides a planet, and forces the planet to move around him in an ellipse. But our sun will also condescend to guide an object no bigger than a shooting star. A bullet, a pea, or even a grain of sand will be held to an elliptic course around the sun as carefully as the great Jupiter himself. The entire shoal of meteors may therefore pursue their common journey around the sun as if inspired by a common purpose, each individual member of the host being, however, guided by the sun, and performing its path in real independence of its neighbours. The orbit followed by this shoal of meteors is enormously large and wide. Here is a sketch of the path (Fig. 77), and I have laid down the position of the orbit of the earth, but not on the same scale. The ellipse is elongated, so that while the shoal approaches comparatively close to the sun at one end of its journey, at the other end it goes out to an enormous distance, far beyond the orbit of the earth—beyond, indeed, the orbit of Jupiter or Saturn; in fact, it reaches to the path of Uranus. To accomplish so vast a journey as this thirty-three years and a quarter are required, and now you will be easily able to see why we get periodical visits from the shoal.

It is, however, a mere piece of good fortune that we

ever encounter the November meteors. Probably there are numerous other shoals of meteors quite as important which we never see, just in the same way as there are many shoals of fish in the sea that never come into our net. The earth moves round the sun in a path which is very nearly a circle, and the shoal moves round in this long oval. We cannot easily represent the true state of things by mere diagrams which

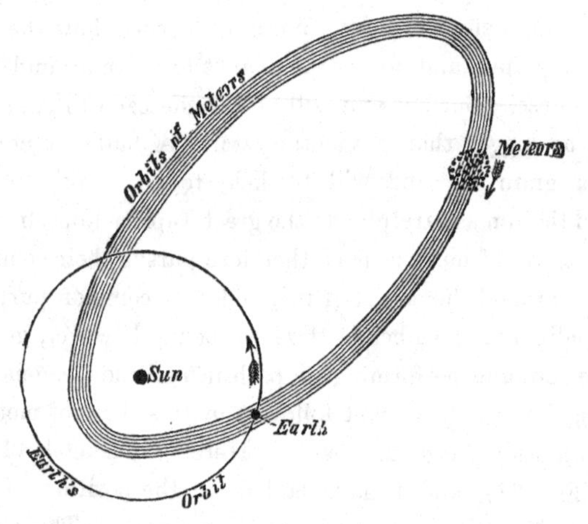

Fig. 78.—The Earth crossing the track of Meteors.

show all these objects on the same plane. This does not give a true representation of the orbits. I think you will better understand what I mean by means of some wire rings. Make a round one to represent the path of the earth, and a long oval one to represent the path of the meteors. There is to be a small opening in the circular ring so that we can slip one of the orbits inside the other. If we are to see the meteors, it is of course necessary that they should strike the earth's atmosphere, for they are not visible to us when they lie at a distance like the moon or like the planets. It is necessary that there be a collision between the earth and the shoal of meteors. But there never could be a collision between two trains unless the lines on which these trains run meet each other; therefore, it is necessary that this long ellipse shall actually cross the earth's track: it will not do to have it pass inside like the two links of a chain; our earth would then miss the meteors altogether, and we would never see them. There are very likely many of such shoals of meteors revolving in this way, and thus escaping our notice entirely.

You will also understand why there is no use in looking for these showers except on the 13th of November. On that day, and on that day alone, the earth appears at that particular point of its route where it crosses the track of the shoal. On the 1st of November, for instance, the earth has not yet reached the point where it could meet with these bodies. By the end of November it has passed too far. But even supposing that the earth is crossing the track of the meteors on the 13th of November, it is still possible that only a few, or none at all, shall be seen. The shoal may not happen to be at that spot at the right time. For a display of meteors to occur, it is therefore necessary that the shoal shall happen to be passing this particular stage of its journey on the 13th of November. In 1866 the earth dipped through the shoal and caught a great many of these meteors in its net. For a few hours the earth was engaged in the capture, until it emerged on the other side of the shoal, and the display was at an end.

Sometimes it happens that in two years following each

other, grand showers of meteors are seen. The reason of this is that the shoal is very long and thin, and consequently if the earth passes through the beginning of the shoal one year, it may have returned to the same point next year before the whole length of the shoal has completely passed. In this case we shall have two great showers in consecutive years. Thus a very fine display was seen in America on the proper day in 1867, while many stragglers were also observed during the three subsequent recurrences of the same date.

Whenever the 13th of November comes round we generally meet with at least a few shooting stars, belonging to this same system, and we must explain how this occurs. Suppose there is a race-course which is comparatively small, so that the competitors will have to run a great many times round before the race is over. Let there be a very large number of entries, and let the majority of the competitors be fairly good runners, while a few are exceptionally good with varying degrees of excellence, and a few are exceptionally bad, some being worse than others. The whole group starts together in a cluster at the signal, and perhaps for the first round or two they may keep tolerably well together. It will be noticed that the cluster begins to elongate as one circuit after another is made: the better runners draw out to the front, and the slower runners lag further and further behind; at last it may happen that those at the head will have gained a whole round on those at the tail, while the other runners of varying degrees of speed will be scattered all round the course. The majority of the competitors, if of nearly equal speed, may continue in a pretty dense group.

Precisely similar has been the great celestial race which these meteors are running. They started on their grand career ages ago, and ever since then they have been flying round and round their mighty course. The great majority of the meteors still stay close together, as their pace is nearly uniform. The exceptionally smart ones have shot ahead, the exceptionally slow ones have lagged behind, and thus it happens that, after fifty or more revolutions have been completed, the shape of the original swarm has become somewhat modified. Its length has been drawn out, while the stragglers and the fastest runners have been scattered all round the path. Across this course our earth carries us every 13th of November; there we usually encounter some of the members of this swarm which have strayed from the great host: they flash into the air, and thus it is that some of these bodies are generally seen every November.

During a shooting star shower it is interesting to notice that all the meteors seem to diverge from a single point. In the adjoining figure (Fig. 78) which shows the directions of a number of meteors' tracks, you will notice that every one seems to radiate from a certain point of the sky. In the case of the shower of the 13th of November this point lies in the constellation of Leo. I must refer you to the Appendix for a description of the way to find Leo or the Lion. The radiant point, as we term it, of this system of meteors is there situated. It is true that the meteors themselves do not generally seem to come all the way from this point. It is the directions of their luminous tails produced backward that carries the eye to the radiant (Fig. 78). If a meteor were seen actually at this point it would be certainly coming straight towards us; it would not then appear as a streak of light at all: it would merely seem like a star which suddenly blazed into splendour and then again sank down into invisibility. Every meteor which appeared near this point would be directed very nearly at the observer, and

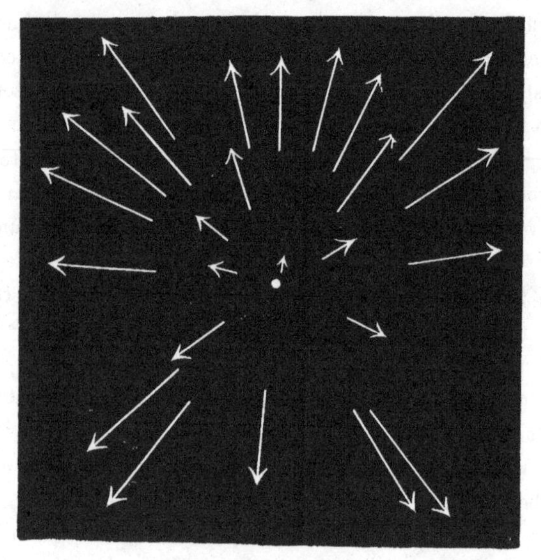

Fig. 79.--The Radiant.

its path would therefore seem very much foreshortened. I can illustrate this with a long straight rod. If I point it directly at you, you can only see the end. If I point it nearly at you, it will seem very much shortened. During the great shower in 1866 I saw many of the meteors so close to the radiant in Leo that they seemed merely like very short marks in the sky; some of them, indeed, seemed to have no length at all. Hence it is that we call this system of shooting stars the "Leonids." They bear this name because their radiant point lies in the constellation Leo, and unless the direction of a shooting star emanates from this point it does not belong to the Leonids. Even if it did so the meteor would not be a Leonid unless the date was right, namely, on the 13th of November, or within a day thereof. We thus have two characters which belong to a shooting star system: there is the date on which they occur and the point from which they radiate.

OTHER GREAT SHOWERS.

To illustrate what I have said, we will speak about another system of shooting stars: they are due every August, from the 9th to the 11th, and their directions diverge from a point in the constellation of Perseus. I may remind you of the recurrence of this shower, as well as that of the November meteors of which we have just spoken, by quoting the following production:—

> "If you November's stars would see,
> From twelfth to fourteenth watching be.
> In August, too, stars shine through heaven,
> On nights between nine and eleven."

It may be worth your while to remember these lines, and always to keep a look-out on the days named. The August meteors, the Perseids we often call them, do not give gorgeous displays, in particular years, with the regularity of the Leonids. There have been, no doubt, some exceptionally grand exhibitions between the 9th and the 11th of August, but we cannot predict when the next is due.

There are vast numbers of stragglers all round the track of the Perseids. In fact, it would seem as if the great race had gone on for such a long period that the cluster had to a great extent broken up, and that a large proportion of the meteors were now scattered the whole way around the course with tolerable uniformity. This being so, it follows that every time we cross the track we are nearly certain to fall in with a few of the stragglers, though we may never enjoy the tremendous spectacle of a plunge through a dense host of meteoroids.

There are many other showers besides the two I have mentioned. Some shooting stars are to be seen every fine night, and those astronomers who pay particular attention to this subject are able to make out scores of small showers of no special beauty. Each of these is fully defined by the night of the year on which it occurs and the position of the point in the heavens from which the meteors radiate. Of all the showers I am only going to mention one more; it is not usually very attractive, but it has a particular interest, as I shall now explain.

It was on the 27th of November, 1872, that a beautiful meteoric shower took place. You will notice, that though the month is the same, the day is entirely different from that on which the Leonids appear. This shower of the 27th is called the Andromedes, because the lines of direction of the shooting stars of which it is composed seem to diverge from a point in the constellation of Andromeda. Ordinarily speaking, there is no special display of meteors connected with the annual return of this day; but in 1872 astronomers were astonished by an exhibition of shooting stars belonging to this system. They were not at all bright when compared with the Leonid meteors. They were, however, quite numerous enough to arrest the attention of very many, even among those who do not usually pay much attention to the heavens.

The chief interest of the shower of Andromedes is in connection with a remarkable discovery connecting meteors and comets. There is a comet which was discovered by the astronomer Biela. It is a small object, requiring a telescope to show it. Biela's comet completes each revolution in a period of about seven years; or rather, I should say, that was the time which the comet used to spend on its journey, for a life of trouble and disaster seems of late to have nearly extinguished the unfortunate object. In 1872 the comet was due in our neighbourhood, and on the night of the 27th of November, in the same year, the earth crossed the track, and, in doing so, the shower of shooting stars was seen. Was this only a coincidence? We crossed the path of the comet at the time when we knew the comet ought to be there; and though we did not then see the comet, we saw a shower of shooting stars, and a wonderful shower too. Does not this seem a remarkable circumstance? Could the comet and the shooting stars be connected together? This is a suggestion we can test in another manner. We know the history of the comet, and we are aware that at the very time of the shower, the comet was approaching us from the direction of the constellation of Andromeda. That is from the very quarter whence the shooting stars have themselves travelled. Taking all these things together, it seems impossible to doubt that the shoal of shooting stars was, if not actually the comet itself, something very closely connected with it.

METEORITES.

Some years ago, a farmer living near Rowton, in Shropshire, noticed on a path in a field a hole which had been suddenly made by some mysterious and unknown agent. The labourers who were near, told him they had just heard a remarkable noise or explosion; and when the farmer put his hand down into the hole, he felt something hot at the bottom of it. He took a spade and dug up the strange body, and found it to be a piece of iron, weighing about seven pounds. He was naturally amazed at such an occurrence, and brought the body home with him.

Where did that piece of iron come from? It is plain that it could not have been always in the ground. The noise and the recently made hole showed that was not the case, and it is confirmed by the fact that the iron was hot. A piece of iron within a few feet of the earth's surface cannot have remained warm for any length of time. It is therefore clear that the iron must have tumbled from the sky. This is a marvellous notion; in fact, it seems so incredible that at first people refused to believe that such things as stones or solid lumps of iron could ever have fallen from the heavens to the earth. But they had to believe it; the evidence was too conclusive. Fortunately, however, the occurrence is a comparatively rare one; indeed, our life on this globe would have an intolerable anxiety added to it if showers of iron hailstones like that at Rowton were at all of frequent occurrence. We should want umbrellas of a more substantial description than those which suffice for the gentle rains we actually experience. There are, indeed, instances on record of persons having been killed by the fearful blows given by these bodies in falling.

But even this lump is a comparatively small one; pieces weighing hundredweights, and even tons, have been collected together in our museums. I would recommend you to pay a visit to that interesting room in our great British Museum in which these meteorites are exhibited. There we have actual specimens of celestial bodies which we can feel and weigh, and which our chemists can analyse. It may be noticed that they only contain substances that we already know on this earth. This celestial iron has often been made use of in primitive times before man understood how to smelt iron from its ore and how to transform it from cast iron to wrought iron. Nature seems to have taken heed of their wants, and occasionally to have thrown down a lump or two for the benefit of those who were so fortunate as to secure them.

That these stones or irons drop from the sky is absolutely certain, but when we try to find out their earlier history we become involved in great difficulties. Nobody really knows, though everybody has a theory on the subject, and thinks, of course, that he must be right and all the others wrong. I have a theory which is, I believe, as likely to be true as any of them, and I call it the Columbiad Theory. I use this expression because every boy or girl listening to me ought to have read Jules Verne's wonderful book, "From the Earth to the Moon," and if any of you have not read it, the sooner you do so the better. It is there narrated how the gun club of Baltimore designed a magnificent cannon which was sunk deep into the ground, and then received a terrific charge of gun cotton, on which a great hollow projectile was carefully lowered, containing inside the three adventurous explorers who desired to visit the moon. Calculations were produced with a view of showing that by firing on a particular day the explosion would drive the projectile up to the moon. There was, however, the necessary condition, that the speed of projection should be great enough. The gun club were accurate in saying that if the cannon were able to discharge the projectile with a speed twenty or thirty times as great as that which had ever been obtained with any other cannon, then the missile would ascend up and up for ever if no further influence were exerted on it. No doubt we have to overlook the resistance of the air and a few other little difficulties, but to this extent, at all events, the gun club were right: that a velocity of about six or seven miles a second would suffice to carry a body away from the gravitation of the earth.

No one supposes that there were ever Columbiad cannons on our globe by which projectiles were shot up into space; but it seems possible that there may have been in very ancient days volcanoes on the earth with a shooting power as great as that which President Barbicane designed for the great cannon.

Even now we have some active volcanoes of no little energy on our earth, and we know that in former days the volcanoes must have been still more powerful; that, in fact, the Vesuvius of the present must be merely a popgun in comparison with volcanoes which have shaken the earth in those primitive days when it had just cooled down from its original fiery condition. It seems not impossible that some of these early volcanoes may in the throes of their mighty eruptions have driven up pieces of iron and volcanic substances with a violence great enough to shoot them off into space.

Suppose that a missile were shot upwards, it would ascend higher and higher, and gravity would, of course, tend to drag it back again down to earth. It can be shown that with an initial speed of six or seven miles a second the missiles would never return to the earth if only dependent on its attraction. The subsequent history of such a projectile would be guided by the laws according to which a planet moves. The body is understood to escape the destination which was aimed at by the Columbiad. I mean, of course, that it is not supposed to hit the moon. Of course, this might conceivably happen; but most of the projectiles would go quite wide of this mark, and would travel off into space.

Though the earth would be unable to recall the projectile, the attraction of the sun would still guide it, whether it was as big as a paving-stone or ever so much larger or smaller. The body would be constrained to follow a path like a little planet around the sun. This track it would steadily pursue for ages. The wanderer would, however, cross the earth's track once during each of its revolutions at the point from which it was projected. Of course it will generally happen that the earth will not be there at the time the meteorite is crossing, and the meteorite will not be there at the time the earth is crossing. Nothing will therefore happen, and the object goes again on its long rounds. But sometimes it must occur that a meteor does not get past the junction without a collision; it plunges into the air, often producing a noise and generating a streak of light like a shooting star. Then it tumbles down, and is restored to that earth whence it originally came.

If this be the true view—and I think there are less weighty objections to it than to any other I know of—then the history of the piece of iron that was found in Shropshire would be somewhat as follows. Many millions of years ago, when the fires of our earth were much more vigorous than they are in these dull times, a terrific volcanic outbreak took place, and vast quantities of material were shot into space, of which this is one of the fragments. During all the ages that have since elapsed this piece of iron has followed its lonely track. In a thousandth part of the time rust and decay would have destroyed it had it lain on the earth, but in the solitudes of space there was found no air or damp to corrode it. At last, after the completion of its long travels, it again crashed down on the earth whence it came.

We have now briefly surveyed the extent of the solar system. We began with the sun which presides over all, and then we discussed the various planets with their satellites, next we considered the eccentric comets, and finally the minute bodies which, as shooting stars or meteorites, must be regarded as forming part of the Sun system. In our closing lecture we shall have to deal with objects of a far more magnificent character.

LECTURE VI.

STARS.

We Try to Make a Map—The Stars are Suns—The Numbers of the Stars —The Clusters of Stars—The Rank of the Earth as a Globe in Space —The Distances of the Stars—The Brightness and Colour of Stars — Double Stars — How we find what the Stars are Made of—The Nebulæ—What the Nebulæ are Made of—Photographing the Nebulæ —Conclusion.

WE TRY TO MAKE A MAP.

THE group of bodies which cluster around our sun forms a little island, so to speak, in the extent of infinite space. We may illustrate this by a map in which we shall endeavour to show the stars placed at their proper relative distances. We first open the compasses one inch, and thus draw a little circle, which I intend to represent the path followed by our earth, the sun being at the centre of the circle. We are not going to put in all the planets. We take Neptune, the outermost, at once. To draw its path I open the compasses to thirty inches and draw a circle with that radius. That will do for our solar system, though the comets no doubt will roam beyond these limits. To complete our map we ought of course to put in some stars. There are a hundred million to choose from, and we shall begin with the brightest. It is often called the Dog star, but astronomers know it better as Sirius. Let us see where it is to be placed on our map. Sirius is beyond Neptune, so it must be outside somewhere. Indeed, it is a good deal further off than Neptune; so I try at the edge of the drawing-board: I have got a method of making a little calculation that I do not intend to trouble you with, but I can assure you that the results it leads me to are quite correct; they show me that this board is not big enough. We must ask the Royal Institution to provide a larger board in this room. But could a board which was big enough fit into this room? Here, again, I make my little calculations, and I find that the room would not hold a board sufficiently great; in fact, if I put the sun here at one end, with its planets around it, Sirius would be too near if it were at the opposite corner of the room. The board would have to go out through the wall of the theatre, out through London. Indeed, big as London is, it would not be large enough to contain the drawing-board that I should require. it would have to stretch about twenty miles from where we are now assembled. We may therefore dismiss any hope of making a practicable map of our system on this scale if Sirius is to have its proper place. Let us, then, take some other star. We shall naturally try with the nearest of all. It is one that we do not know in this part of the world, but those who live in the southern hemisphere are well acquainted with it. The name of this star is Alpha Centauri. Even for this star, we should require a drawing three or four miles long if the distance from the earth to the sun is to be taken as one inch. You see what an isolated position our sun and his planets occupy. The members of the family are all close together, and the nearest neighbours are situated at enormous distances. There is a good reason for this separation. The stars are very pretty where they lie, but, as they might be very troublesome neighbours if they were close to our system, it is well they are so far off; they would be constantly making disturbance in the sun's family if they were near at hand. Sometimes they would be dragging us into hot water by bringing us too close to the sun, or producing a coolness by pulling us away from the sun, which would be quite as disagreeable.

THE STARS ARE SUNS.

We are about to discuss one of the grandest truths in the whole of nature. We have had occasion to see that this sun of ours is a magnificent globe immensely larger than the greatest of his planets, while the greatest of these planets is immensely larger than this earth; but now we are to learn that our sun is, after all, only a star not nearly so bright as many of those which shine over our heads every night. We are comparatively close to the sun, so that we are able to enjoy his beautiful light and cheering heat. All those other myriads of stars are each of them suns, and the splendour of those distant suns is often far greater than that of our own. We are, however, so enormously far from them that they have dwindled down to insignificance. To judge impartially between our star or sun and such a sun or star as Sirius we should stand half-way between the two: it is impossible to make a fair estimate when we find ourselves situated close up to one star and a million times as far from the other. When we make allowance for the imperfections of our point of view, we are enabled to realise the majestic truth that our sun is no more than a star, and that the other stars are no less than suns.

This gives us an imposing idea of the extent and the magnificence of the universe in which we are situated. Look up to the sky at night—you will see a host of stars: try to think that every one of them is itself a sun. It may probably be that those suns have planets circulating around them, but it is hopeless for us to expect to see such planets. Were you standing on one of those stars and looking towards our system, you would not perceive the sun to be the brilliant and gorgeous object that we know so well. If you could see him at all, he would merely seem like a star, not nearly so bright as many of those you can see at night. Even if you had the biggest of telescopes to aid your vision, you could never discern from one of these bodies the planets which surround the sun. No astronomer in the stars could see Jupiter, even if his sight were a thousand times as good or his telescopes a thousand times as powerful as any sight or telescope that we know. So minute an object as our earth would, of course, be still more hopelessly beyond the possibility of vision.

THE NUMBERS OF THE STARS.

To count the stars involves a task which lies beyond the power of man to fully accomplish. Even without the aid of any telescope, we can see a great multitude of stars from this part of the world. There are also many constellations in the southern hemisphere which never appear above our horizon. If, however, we were to go to the equator, then, by waiting there for a twelvemonth, all the stars in the heavens would have been successively exposed to view. An astronomer, Houzeau, who had the patience to count them, enumerated about 6,000. This is the naked-eye estimate of the star-population of the heavens; but if, instead of relying on unassisted eyes, you get the assistance of a little telescope, you will be astounded at the enormous multitude of stars which are disclosed.

An ordinary opera-glass is a very useful instrument

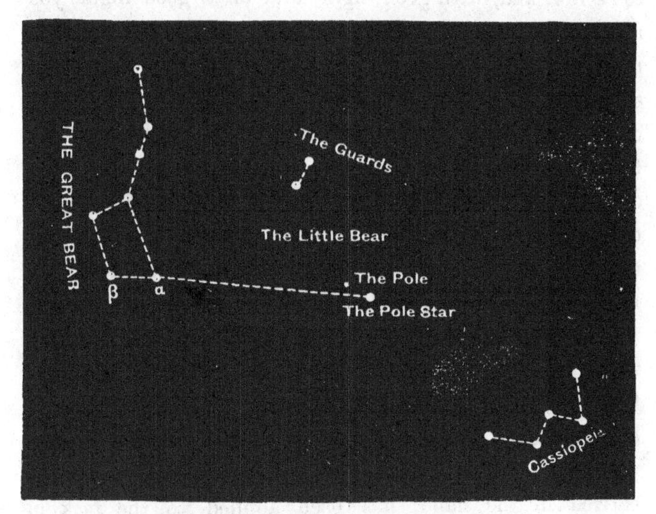

Fig. 80.—The Great Bear and the Pole.

for looking at the stars in the heavens, as well as at the stars of another description, to which it is more commonly applied. You will be amazed to find that the heavens teem with additional multitudes of stars that the opera-glass will reveal. Any part of the sky may be observed; but, just to give an illustration, I shall take one special region, namely, that of the Great Bear (Fig. 80). The seven well-known stars are here shown, four of which form a sort of oblong, while the other three represent the tail. In an Appendix I shall describe the services of the Great Bear as a guide to a knowledge of the constellations. I would like you to make this little experiment. On a fine clear night, count how many stars there are within this oblong; they are all very faint, but you will be able to see a few, and, with good sight, and on a good night, you may see perhaps ten. Next take your opera-glass and sweep it over the same region: if you will carefully count the stars it shows you will find fully 200; so that the opera-glass has, in this part of the sky, revealed nearly twenty times as many stars as could be seen without its aid. As 6,000 stars can be seen by the eye, all over the heavens, we may fairly expect that twenty times that number—that is to say, 120,000 stars—could be shown by the opera-glass over the entire sky. Let us go a step further, and employ a telescope, the object-glass of which is three inches across. This is a useful telescope to have, and, if a good one, will show multitudes of pleasing objects, though an astronomer would not consider it very powerful. An instrument like this, small enough to be carried in the hand, has been applied to the task of enumerating the stars in the northern half of the sky, and 320,000 stars were counted. Indeed, the actual number that might have been seen with it is considerably greater, for when the astronomer Argelander made this memorable investigation he was unable to reckon many of the stars in localities where they lay very close together. This grand count only extended to half the sky, and, assuming that the other half is as populous, we see that a little telescope like that we have supposed will show over the sky a number of stars which exceeds that of the population of any city in England except London. It exhibits more than one hundred times as many stars as our eyes could possibly reveal. Still, we are only at the beginning of the count: the really great telescopes add largely to the number. There are multitudes of stars which, in small telescopes, we cannot see, but which are distinctly visible from our great observatories. That telescope would be still but a comparatively small one which would show as many stars in the sky as there are people living in this mighty city of London; and with the greatest instruments, the tale of stars has risen to a number far greater than that of the entire population of Great Britain.

In addition to those stars the largest telescopes show us, there are myriads which make their presence evident in a wholly different way. It is only in quite recent times that an attempt has been made to develop fully the powers of photography in representing the celestial objects. On a photographic plate which has been exposed to the sky in a great telescope the stars are recorded in their thousands. Many of these may, of course, be observed with a good telescope, but there are not a few others which no one ever saw in a telescope, which apparently no one ever could see, though the photograph is able to show them. We do not, however, employ a camera like that which the photographer uses who is going to take your portrait. The astronomer's plate is put into his telescope, and then the telescope is turned towards the sky. On that plate the stars produce their images, each with its own light. Some of these images are excessively faint, but we give a very long exposure of an hour or two hours; sometimes so much as four hours' exposure is given to a plate so sensitive that a mere fraction of a second would sufficiently expose it during the ordinary practice of taking a photograph in daylight. We thus afford sufficient time to enable the fainter objects to indicate their presence upon the sensitive film. Even with an exposure of a single hour a picture exhibiting 16,000 stars has been taken by Mr. Isaac Roberts, of Liverpool. Yet the portion of the sky which it represents is only one ten-thousandth part of the entire heavens. It should be added that the region which Mr. Roberts has photographed is furnished with stars in rather exceptional profusion.

Here, at last, we have obtained some conception of the sublime scale on which the stellar universe is constructed. Yet even these plates cannot represent all the stars that the heavens contain. We have every reason for knowing that with larger telescopes, with more sensitive plates, with more prolonged exposures, ever fresh myriads of stars will be brought within our view.

You must remember that every one of these stars is truly a sun, a lamp as it were, which doubtless gives light to other objects in its neighbourhood as our sun sheds light upon this earth and the other planets. In fact, to realise the glories of the heavens you should try to think that the brilliant points you see are merely the luminous points of the otherwise invisible universe.

Standing one fine night on the deck of a Cunarder we passed in open ocean another great Atlantic steamer. The vessel was near enough for us to see not only the light from the mast-head but also the little beams from the several cabin ports; but we could see nothing of the ship herself.

Her very existence was only known to us by the twinkle of these lights. Doubtless her passengers could see, and did see, the similar lights from our own vessel, and they doubtless drew the correct inference that these lights indicated a great ship.

Consider the multiplicity of beings and objects in a ship: the captain and the crew, the passengers, the cabins, the engines, the boats, the rigging, and the stores. Think of all the varied interests there collected and then reflect that out on the ocean, at night, the sole indication of the existence of this elaborate structure was given by the few beams of light that happened to radiate from it. Now raise your eyes to the stars, there are the twinkling lights. We cannot see what those lights illuminate, nor can we conjecture what untold wealth of non-luminous bodies may also lie in their vicinity; we may, however, feel certain that just as the few gleaming lights from a ship are utterly inadequate to give a notion of the nature and the contents of an Atlantic steamer, so are the twinkling stars utterly inadequate to give even the faintest conception of the extent and the interest of the universe. We merely see self-luminous bodies, but of the multitudes of objects and the elaborate systems of which these bodies are only the conspicuous points we see nothing and we know nothing. We are, however, entitled to infer from an examination of our own star—the sun—and of the beautiful system by which it is surrounded, that these other suns may be also splendidly attended. This is quite as reasonable a supposition as that a set of lights seen at night on the Atlantic Ocean indicate the existence of a fine ship.

THE CLUSTERS OF STARS.

On a clear night you can often see, stretching across the sky, a track of faint light, which is known to astronomers as the "Milky Way." It extends below the horizon and then round the earth to form a girdle about the heavens. When we examine the Milky Way with a telescope we find, to our amazement, that it consists of myriads of stars, so small and so faint that we are not able to distinguish them individually, we merely see the glow produced from their collective rays. Remembering that our sun is a star, and that the Milky Way surrounds us, it would almost seem as if our sun were but one of the host of stars which form this cluster.

There are also other clusters of stars, some of which are most exquisitely beautiful telescopic spectacles. I may mention a celebrated pair of these objects which lie in the constellation of Perseus. The sight of these in a great telescope is so imposing that no one who is fit to look through a telescope could resist a shout of wonder and admiration when first they burst on his view. But there are other clusters. Here is a picture of one which is known as the "Globular Cluster in the Centaur" (Fig. 81). It consists of a ball of stars, so far off, that however large these several suns may actually be, they have dwindled down to extremely small points of light. A homely illustration may serve to show the appearance which a globular cluster presents in a good telescope. I take a pepper-castor and on a sheet of white paper I begin to shake out the pepper until there is a little heap at the centre and other grains are scattered loosely about. Imagine that every one of those grains of pepper was to be transformed into a tiny electric light, and then you have some idea of what a cluster of stars would look like when viewed through a telescope of sufficient power. There are multitudes of such clusters scattered through the depths of space. They require our biggest telescopes to show them adequately. We have seen that our sun is a star, being only one of a magnificent cluster that form the Milky Way. We have also seen that there

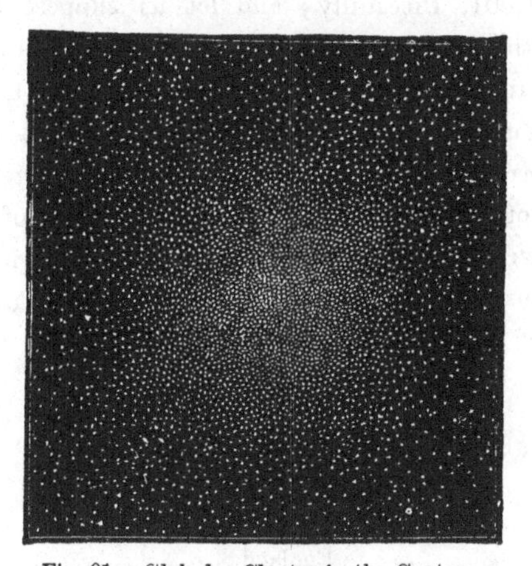

Fig. 81.—Globular Cluster in the Centaur.

are other clusters scattered through the length and depth of space. It is thus we obtain a notion of the rank which our earth holds in the scheme of things celestial.

THE RANK OF THE EARTH AS A GLOBE IN SPACE.

Let me give an illustration with the view of explaining more fully the nature of the relation which the earth bears to the other globes which abound through space, and you must allow me to draw a little upon my imagination.

I shall suppose that Her Majesty's mails extend not only over this globe, but that they also communicate with other worlds; that postal arrangements exist between Mars and the earth, between the sun and Orion—in fact, everywhere throughout the whole extent of the universe. We shall consider how our letters are to be addressed. Let us take the case of Mr. John Smith, merchant, who lives at 1,001, Piccadilly; and let us suppose that Mr. John Smith's business transactions are of such an extensive nature that they reach not only all over this globe, but away throughout space. I shall suppose that the firm has a correspondent residing — let us say in the constellation of the Great Bear; and when this man of business wants to write to Mr. Smith from these remote regions, what address must he put upon the letter, so that the Postmaster-General of the universe shall make no mistake about its delivery? He will write as follows:—

MR. JOHN SMITH,
1,001, Piccadilly,
London,
England,
Europe,
Earth,
Near the Sun,
Milky Way,
The Universe.

Let us now see what the several lines of this address mean. Of course we put down the name of Mr. John Smith in the first line, and then we will add "1 001, Piccadilly," for the second; but as the people in the Great Bear are not likely to know where Piccadilly is, we shall add "London" underneath. As even London itself cannot be well known everywhere, it is better to write "England" underneath. This would surely find Mr. John Smith from any post-office on this globe. From other globes, however, the supreme importance of England may not be so immediately recognised, and therefore it is as well to add another line "Europe." This ought to be sufficient, I think, for any post-office in the solar system. Europe is big enough to be visible from Mars or Venus, and should be known to the post-office people there, just as we know and have names for the continents on Mars. But further away there might be a little difficulty: from Uranus and Neptune the different regions on our earth can never have been distinguished, and therefore we must add another line to indicate the particular globe of the solar system which contains Europe. Mark Twain tells us that there was always one thing in astronomy which specially puzzled him, and that was to know how we found out the names of the stars. We are, of course, in hopeless ignorance of the name by which this earth is called among other intelligent beings elsewhere who can see it. I can only adopt the title of "Earth," and therefore I add this line. Now our address is so complete that from anywhere in the solar system—from Mercury, from Jupiter, or Neptune—there ought to be no mistake about the letter finding its way to Mr. John Smith. But from his correspondent in the Great Bear this address would be still incomplete; they cannot see our earth from thence, and even the sun himself only looks like a small star—like one, in fact, of thousands of stars elsewhere. However, each star can be distinguished, and our sun may, for instance, be recognised from the Great Bear by some designation.

We shall add the line "Near the Sun," and then I think that from this constellation, or from any of the other stars around us, the address of Mr. John Smith may be regarded as complete. But Mr. Smith's correspondence may be still wider. He may have an agent living in the cluster of Perseus or on some other objects still fainter and more distant; then "Near the Sun" is utterly inadequate as a concluding line to the address, for the sun, if it can be seen at all from thence, will be only of the significance of an excessively minute star, no more to be designated by a special name than are the several leaves on the trees of a forest. What this distant correspondent will be acquainted with is not the earth or the sun, but only the cluster of stars among which the sun is but a unit. Again we use our own name to denote the cluster, and we call it the "Milky Way." When we add this line, we have made the address of Mr. John Smith as complete as circumstances permit it to be. I think a letter posted to him anywhere ought to reach its destination. For completeness, however, we will finish up with one line more—"*The Universe.*"

THE DISTANCES OF THE STARS.

I must now tell you something about the distances of the stars. I shall not make the attempt to explain fully how astronomers make such measurements, but I will give you some notion of how it is done. You may remember I showed you how we found the distance of a globe that was hung from the ceiling. The principle of the method for finding the distance of the star is somewhat similar, except that we make the two observations not from the two ends of a table, not even from the two sides of the earth, but from two opposite points on the earth's orbit, which are therefore at a distance of 186,000,000 miles. Imagine that on Midsummer Day, when standing on the earth here, I measure with a piece of card the angle between the star and the sun. Six months later, on Midwinter Day, when the earth is at the opposite point of its orbit, I again measure the angle between the same star and the sun, and we can now determine the star's distance by making a triangle. I draw a line a foot long, and we will take this foot to represent 186,000,000 miles, the distance between the two stations; then, placing the cards at the corners, I rule the two sides and complete the triangle, and the star must be at the remaining corner; then I measure the sides of the triangle and find how many feet they contain, and recollecting that each foot corresponds to 186,000,000 miles, we discover the distance of the star. If the stars were comparatively near us, the process would be a very simple one; but, unfortunately, the stars are so extremely far off, that this triangle, even with a base of only one foot, must have its sides many miles long. Indeed, astronomers will tell you that there is no more delicate or troublesome work in the whole of their science than that of discovering the distance of a star.

In all such measurements we take the distance from the earth to the sun as a conveniently long measuring rod, whereby to express the results. The nearest stars are still hundreds of thousands of times as far off as the sun. Let us ponder for a little on magnitudes so vast. We shall first express them in miles. Taking the sun's distance to be 93,000,000 miles, then the distance of the nearest fixed star is about twenty millions of millions of miles—that is to say, we express this distance by putting down a 2 first, and then writing thirteen cyphers after it. It is, no doubt, easy to speak of such figures, but it is a very different matter when we endeavour to imagine the awful magnitude which such a number indicates. I must try to give some illustrations which will enable you to form a notion of it. At first I was going to ask you to try and count this number, but when I found it would require at least 300,000 years, counting day and night without stopping before the task was over, it became necessary to adopt some other method.

When lately in Lancashire I was kindly permitted to visit a cotton mill, and I learned that the cotton yarn there produced in a single day would be long enough to wind round this earth twenty-seven times at the equator. It appears that the total production of cotton yarn each day in all the mills together would be on the average about 155,000,000 miles. In fact, if they would only spin about one-fifth more, we could assert that Great Britain produced enough cotton yarn every day to stretch from the earth to the sun and back again! It is not hard to find from these figures how long it would take for all the mills in Lancashire to produce a piece of yarn long enough to reach from our earth to the nearest of the stars. If the spinners worked as hard as ever they could for a year, and if all the pieces were then tied together, they would extend to only a small fraction of the distance; nor if they worked for ten years, or for twenty years, would the task be fully accomplished. Indeed, upwards of 400 years would be necessary before enough cotton could be grown in America and spun in this country to stretch over a distance so enormous. All the spinning that has ever yet been done in the world has not formed a long enough thread!

There is another way in which we can form some notion of the immensity of these sidereal distances. You will recollect that, when we were speaking of Jupiter's moons (p. 205), I told you of the beautiful discovery which their eclipses enabled astronomers to make. It was thus found that light travels at the enormous speed of about 185,000 miles per second. It moves so quickly that within a single second a ray would flash two hundred times from London to Edinburgh and back again.

We said that a meteor travels one hundred times as swiftly as a rifle-bullet; but even this great speed seems almost nothing when compared with the speed of light, which is 10,000 times as great. Suppose some brilliant outbreak of light were to take place in a distant star—an outbreak which would be of such intensity that the flash from it would extend far and wide throughout the universe. The light would start forth on its voyage with terrific speed. Any neighbouring star which was at a distance of less than 185,000 miles would, of course, see the flash within a second after it had been produced. More distant bodies would receive the intimation after intervals of time proportional to their distances. Thus, if a body were 1,000,000 miles away the light would reach it in five or six seconds, while over a distance as great as that which separates the earth from the sun the news would be carried in about eight minutes. We can calculate how long a time must elapse ere the light shall travel over a distance so great as that between the star and our earth. You will find that from the nearest of the stars the time required for the journey will be over three years. Ponder on all that this involves. That outbreak in the star might be great enough to be visible here, but we could never become aware of it till three years after it had happened. When we are looking at such a star to-night we do not see it as it is at present, for the light that is at this moment entering our eyes has travelled so far that it has been three years on the way, therefore, when we look at the star now we see it as it was three years previously. In fact, if the star was to go out altogether, we might still continue to see it twinkling away for a period of three years longer, because a certain amount of light was on its way to us at the moment of extinction, and so long as that light keeps arriving here, so long shall we see the star showing as brightly as ever. When, therefore, you look at the thousands of stars in the sky to-night, there is not one that you see as it is now, but as it was years ago.

I have been speaking of the stars that are nearest to us, but there are others much farther off. It is true we cannot find the distance of these more remote objects with any degree of accuracy, but we can convince ourselves how great that distance is by the following reasoning. Look at one of the brightest stars. Try to conceive that the object was drawn away further into the depths of space, until it was ten times as far from us as it is at present, it would still remain bright enough to be recognised in quite a small telescope; even if it were taken to one hundred times its original distance it would not have withdrawn from the view of a good telescope; while if it retreated one thousand times as far as it was at first it would still be a recognisable point in our mightiest instruments. Among the stars which we can see in our telescopes, we feel confident there must be many from which the light has taken hundreds of years, or even thousands of years, to arrive here. When, therefore, we look at such objects, we see them, not as they are now, but as they were ages ago; in fact, a star might have ceased to exist for thousands of years, and still be seen by us every night as a twinkling point in our great telescopes.

Remembering these facts, you will, I think, look at the heavens with a new interest. There is a bright star, Vega or Alpha Lyræ, a beautiful gem, but so far off that the light from it which we now see started before many of my audience were born. Suppose that there are astronomers residing on worlds amid the stars, and that they have sufficiently powerful telescopes to view this globe, what do you think they will observe? They will not see our earth as it is at present, they will see us as we were years ago. There are stars from which, if England could now be seen, the whole of the country would be observed at this present moment to be in a great state of excitement at a very auspicious event. Distant astronomers might observe a great procession in London amid the enthusiasm of a nation, and they could watch the coronation of a youthful queen. There are other stars still further off, from which, if the inhabitants had good enough telescopes, they would now see a mighty battle in progress not far from Brussels: they would see one army dashing itself time after time against the immovable ranks of the other. I do not think they would be able to hear the ever-memorable, "Up, Guards, and at them!" but there can be no doubt that there are stars so far away that the rays of light which started from the earth on the day of the Battle of Waterloo are only just arriving there. Further off still, there are stars from which a bird's-eye view could be taken at this very moment of the signing of Magna Charta. There are even stars from which England, if it could be seen at all, would now appear, not as the great England we know, but as a country covered by dense forests, and inhabited by painted savages, who waged incessant war with wild beasts that roamed through the island. The geological problems that now puzzle us would be quickly solved could we only go far enough into space and had we only powerful enough telescopes. We should then be able to view our earth through the successive epochs of past geological time: we should be actually able to see those great animals whose fossil remains are treasured in our museums, tramping about over the earth's surface, splashing across its swamps, or swimming with broad flippers through its oceans. Indeed, if we could view our own earth reflected from mirrors in the stars, we could still see Moses crossing the Red Sea, or Adam and Eve being expelled from Eden.

So important is the subject of star distance that I am tempted to give one more illustration in order to bring before you some conception of how vast that distance truly is. I shall take, as before, the nearest of the stars so far as known to us, and I hope to be forgiven for taking an illustration of a practical and a commercial kind instead of one more purely scientific. I shall suppose that a railway is about to be made from London to Alpha Centauri. The length of that railway, of course, we have already stated: it is twenty billions of miles. So I am now going to ask your attention to the simple question as to the fare which it would be reasonable to charge for the journey. We shall choose a very cheap scale on which to compute the fare. The parliamentary rate here is, I believe, a penny for every mile. We will make our interstellar railway fares much less even than this; we shall arrange to travel at the rate of one hundred miles for every penny. That, surely, is moderate enough. If our fares were so low that the journey from London to Edinburgh only cost fourpence, then even the most unreasonable passenger would be surely contented. On these terms how much do you think the fare from London to this star ought to be? I know of one way in which to make the answer intelligible. There is a National Debt with which your fathers are, unhappily, only too well acquainted: you will know quite enough about it yourselves in those days when you have to pay income tax. This Debt is so vast that the interest upon it alone is about sixty thousand pounds a day, the whole amount of the National Debt being seven hundred and thirty-six millions of pounds (April, 1887).

If you went to the booking-office with the whole of this mighty sum in your pocket—but stop a moment: could you carry it in your pocket? Certainly not, if it were in sovereigns. You would find that after your pocket had as many sovereigns as it could conveniently hold there would still be some left—so many, indeed, that it would be necessary to get a cart to help you on with the rest. When the cart had as great a load of sovereigns as the horse could draw there would be still some more, and you would have to get another cart; but ten carts, twenty carts, fifty carts, would not be enough. You would want five thousand carts before you would be able to move off towards the station with your money. When you did get there and asked for a ticket at the rate of 100 miles for a penny, do you think you would get any change back? No doubt some little time would be required to count the money, but when it was counted the clerk would tell you that it was not enough, that he must have nearly a hundred millions of pounds more.

That will give some notion of the distance of the nearest star, and we may multiply it by ten, by one hundred, and even by one thousand, and still not attain to the distance of some of the more distant stars that the telescope shows us.

On account of the immense distances of the stars we can only perceive them to be mere points of light. You will never see a star to be a globe with marks on it like the moon, or like one of the planets—in fact, the better the telescope the smaller does a star seem, though, of course, its brightness is increased with every increase in the light-grasping power of the instrument.

THE BRIGHTNESS AND COLOUR OF STARS.

Another point to be noticed is the arrangement of stars in classes, according to their lustre. The brightest stars, of which there are about twenty, are said to be of the first magnitude. The stars just inferior to the first magnitude are ranked as the second; and those just inferior to the second are estimated as the third; and so on. The smallest stars that your unaided eyes will show you are of about the sixth magnitude. Then the telescope will reveal stars still fainter and fainter, down to what we term the seventeenth or eighteenth magnitudes, or even lower still. The number of stars of each magnitude increases very much in the classes of small stars.

There is one of the larger stars which should be specially mentioned. It is Sirius, or the Dog-star, which is much brighter than any other in the heavens. Most of the stars are white, but many are of a somewhat ruddy hue. There are a few telescopic stars which are intensely red, some exhibit beautiful golden tints, while others are blue or green.

There are some curious stars which regularly change their brilliancy. Let me try to illustrate the nature of these variables. Suppose that you were looking at a street gas-lamp from a very long distance, so that it seemed a little twinkling light; and suppose that someone was attending to turn the cock up and down. Or, better still, imagine a little machine which would act regularly so as to keep the light first of all at its full brightness for two days and a half, and then gradually turn it down until in three or four hours it declines to a feeble glimmer. In this low state the light remains for twenty minutes; then during three or four hours the gas is to be slowly and gradually turned on again until it is full. In this condition the light will remain for two days and a half, and then the same series of changes is to commence. This would be a very odd form of gas-lamp. There would be periods of two days and a half during which it would remain at its full; these would be separated by intervals of about seven hours, when the slow turning down and turning up again would be in progress.

The imaginary gas-lamp is exactly paralleled by a star

Algol, in the constellation of Perseus (Fig. 82), which goes through the series of changes I have indicated. Ordinarily speaking, it is a bright star of the second magnitude, and

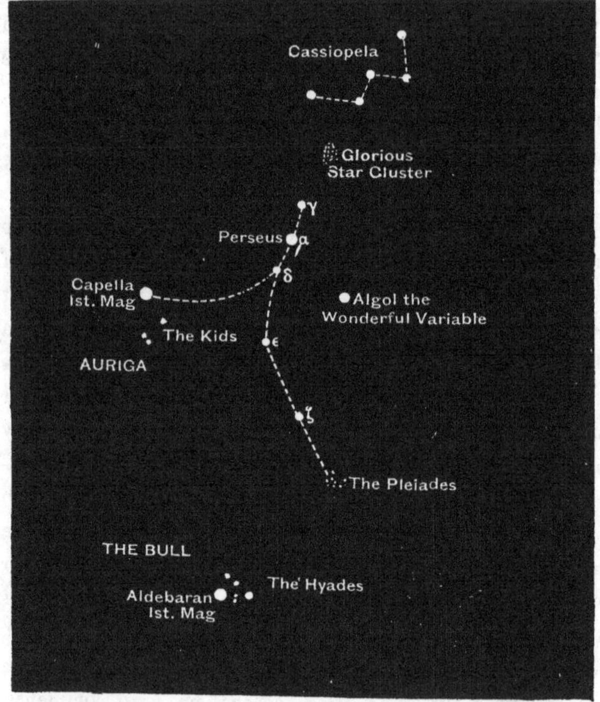

Fig. 82.—Perseus and its neighbouring Stars, including Algol.

whatever be the cause, the star performs its variations with marvellous uniformity. In fact, Algol has always arrested the attention of those who observed the heavens, and in early times was looked on as the eye of a Demon. There are many other stars which also change their brilliancy.

Most of them require much longer periods than Algol, and sometimes a new star which nobody has ever seen before will suddenly kindle into brilliancy.

DOUBLE STARS.

Whenever you have a chance of looking at the heavens through a telescope, you should ask to be shown what is called *a double star*. There are many stars in the heavens which present no remarkable appearance to the unaided eye, but which a good telescope at once shows to be of quite a complex nature. These are what we call double stars, in which the two little points of light are placed so close together that the unaided eye is unable to separate them. Under the magnifying power of the telescope, however, they are seen to be distinct. In order to give some notion of what these objects are like, I shall briefly describe three of them. The first lies in that best known of constellations, the Great Bear. If you look at his tail, which consists of three stars, you will see that near the middle one of the three a small star is situated; we call this star Alcor, but it is the brighter one near Alcor to which I specially call your attention. The sharpest eye would never suspect that this object was composed of two stars placed close together. Even a small telescope will, however, show this to be the case, and this is the easiest and the first observation that a young astronomer should make when beginning to turn a telescope to the heavens. Of course you will not imagine that I mean Alcor to be the second component of the double star; it is the bright star near Alcor which is the double. Here are two marbles, and these marbles are fastened an inch apart. You can see them, of course, to be separate; but if the pair were moved further and further away, then you would soon not be able to distinguish between them, though the actual distance between the marbles had not altered. Look at these two wax tapers which are now lighted; the little flames are an inch apart. You would have to view them from a station a third of a mile away if the distance between the two flames were to appear the same as that between the two components of this double star. Your eye would never be able to discriminate between two lights only an inch apart at so great a distance; a telescope would, however, enable you to do so, and this is the reason why we have to use telescopes to show us double stars.

You might look at that double star year after year throughout the course of a long life without finding any appreciable change in the relative positions of its components. But we know that there is no such thing as rest in the universe; even if you could balance a body so as to leave it for a moment at rest, it would not stay there, for the simple reason that all the bodies around it in every direction are pulling at it, and it is certain that the pull in one direction will preponderate so that move it must. Especially is this true in the case of two suns like those forming a double star. Placed comparatively near each other they could not remain permanently in that position; they must gradually draw together and come into collision with an awful crash. There is only one way by which such a disaster could be obviated. That is by making one of these stars revolve around the other just as the earth revolves around the sun, or the moon revolves around the earth. There must, therefore, be some motion going on in every genuine double star, whether we have been able to see that motion or not.

Let us now look at another double star of a different kind. This time it is in the constellation of Gemini. The heavenly twins are called Castor and Pollux. Of these, Castor is a very beautiful double star, consisting of two bright points, a great deal closer together than were those in the Great Bear; consequently a better telescope is required for the purpose of showing them separately. Castor has been watched for many years, and it can be seen that one of these stars is slowly revolving around the other; but it takes a very long time, amounting to hundreds of years, for a complete circuit to be accomplished. This seems very astonishing, but when you remember how exceedingly far Castor is from us, you will see that that pair of stars which seems so close together that it requires a telescope to show that they are distinct must indeed be separated by hundreds of millions of miles. Let us try to conceive our own system transformed into a double star. If we took our outermost planet—Neptune—and enlarged him a good deal, and then heated him sufficiently to make him glow like a sun, he would still continue to revolve round our sun at the same distance, and thus a double star would be produced. An inhabitant of Castor who turned his telescope towards us, would be able to see the sun as a star. He would not, of course, be able to see the earth, but he might see Neptune like another small star close up to the sun. If generations of astronomers in Castor continued their observations of our system, they would find a binary star, of which one component took a century and a half to go round the other. Need we then be surprised that when we look at Castor we observe movements equally deliberate?

There is often so much glare from the bright stars seen

in a telescope, and so much twinkling in some states of the atmosphere, that stars appear to dance about in rather a puzzling fashion, especially to one who is not accustomed to astronomical observations. I remember hearing how a gentleman once came to visit an observatory. The astronomer showed him Castor through a powerful telescope as a fine specimen of a double star, and then, by way of improving his little lesson, the astronomer mentioned that one of these stars was revolving around the other. "Oh, yes," said the visitor, "I saw them going round and round in the telescope." He would, however, have had to wait for a few centuries with his eye to the instrument before he would have been entitled to make this assertion.

Double stars also frequently delight us by giving beautifully contrasted colours. I dare say you have often noticed the red and the green lights that are used on railways in the signal lamps. Imagine one of those red and one of those green lights away far up in the sky and placed close together, then you would have some idea of the appear ance that a coloured double star presents, though, perhaps, I should add that the hues in the heavenly bodies are not so vividly contrasted as are those which our railway people find necessary. There is a particularly beautiful double star of this kind in the constellation of the Swan. You could make an imitation of it by boring two holes, with a red hot needle, in a piece of card, and then covering one of these holes with a small bit of the topaz-coloured gelatine with which Christmas crackers are made. The other star is to be similarly covered with blue gelatine. A slide made on this principle placed in the lantern gives a very good representation of these two stars on the screen. There are many other coloured doubles besides this one; and, indeed, it is noteworthy that we hardly ever find a blue or a green star by itself in the sky: it is always as a member of one of these pairs.

HOW WE FIND WHAT THE STARS ARE MADE OF.

Here is a piece of stone. If I wanted to know what this stone was composed of, I should ask a chemist to tell me. He would take it into his laboratory, and first crush it up into powder, and then, with his test tubes, and with the liquids which his bottles containt and his weighing scales, and other apparatus, he will, tell all about it: there is so much of this, and so much of that, and plenty of this, and none at all of that. But now, suppose you ask this chemist to tell you what the sun is made of, or one of the stars. Of course, you have not a sample of it to give him; how then, can he possibly find out anything about it? Well, he can find out something, and this is the wonderful discovery that I want to explain to you. We now put down the gas, and I kindle a brilliant red light. Perhaps some of those whom I see before me have occasionally ventured on the somewhat dangerous practice of making fireworks. If there is any boy here who has ever made sky-rockets, and put the little balls into the top, which are to burn with such vivid colours when the explosion takes place, he will know that the substance which tinged that red fire must have been *strontium.* He will recognise it by the colour; because *strontium* gives that red light which nothing else will give. Here are some of these lightning papers, as they are called; they are very pretty and very harmless; and these, too, give brilliant red flashes as

I throw them. The red tint has, no doubt, been given by *strontium* also. You see we recognised the substance simply by the colour of the light it produced when burning.

Perhaps some of you have tried to make a ghost at Christmas time by dressing up in a sheet, and bearing in your hand a ladle blazing with a mixture of common salt and spirits of wine, the effect produced being a most ghostly one. Some mammas will hardly thank me for this suggestion, unless I add that the ghost must walk about cautiously, for otherwise the blazing spirit would be very apt to produce conflagrations of a kind quite different from those intended. However, by the kindness of Professor Dewar, I am enabled to show the phenomenon on a splendid scale, and also free from all danger. I kindle a vivid flame of an intensely yellow colour, which I think the ladies will unanimously agree is not at all becoming to their complexions, while the pretty dresses have lost all their variety of colours. Here is a nice bouquet, and yet you can hardly distinguish the green of the leaves from the brilliant colours of the flowers, except by trifling difference of shade. Expose to this light a number of pieces of variously coloured ribbon, pink and red and green and blue, and all their beauty is gone; and yet we are told that this yellow is a perfectly pure colour; in fact the purest colour that can be produced. I think we have to be thankful that the light which our good sun sends us does not possess purity of that description. There is just one substance which will produce that yellow light; it is a curious metal called sodium—a metal so soft that you can cut it with a knife, and so light that it will float on water;

while, still more strange, it actually takes fire the moment it is dropped on the water. It is only in a chemical laboratory that you will be likely to meet with the actual metallic sodium, yet in other forms the substance is one of the most abundant in nature. Indeed, common salt is nothing but sodium closely united with a most poisonous gas, a few respirations of which would kill you. But you see this strange metal and this noxious gas, when united, become simply the salt for our eggs at breakfast. This simple yellow light, wherever it is seen, either in the flame of spirits of wine mixed with salt or in that great blaze at which we have been looking, is a characteristic of sodium. Wherever you see that particular kind of light, you know that sodium must have been present in the body from which it came. We have accordingly learned to recognise two substances, namely, *strontium* and *sodium*, merely by the light which they give out when heated so as to burn. To these we may add a third. Here is a strip of white metallic ribbon. It is called magnesium. It seems like a bit of tin at the first glance, but, indeed, it is a very different thing from tin; for, look, when I hold it in the spirit-lamp, the strip of metal immediately takes fire, and burns with a white light so dazzling that it pales the gas-flames to insignificance. There is no other substance that will, when kindled, give that particular kind of light which we see from magnesium. I can recommend this little experiment as quite suitable for trying at home; you can buy a bit of magnesium ribbon for a trifle at the optician's; it cannot explode or do any harm, nor will you get into any trouble with the authorities provided you hold it when burning over a tray or a newspaper, so as to prevent the white ashes from falling on the carpet.

There are, in nature, a number of simple substances called elements. Every one of these, when ignited under suitable conditions, emits a light which belongs to it alone and by which it can be distinguished from every other substance. I do not say that we can try the experiments in the simple way I have here indicated. Many of the substances will only yield their light so as to be recognisable by much more elaborate artifices than those which have sufficed for us. But you see that the method affords a means of finding out the actual substances present in the sun or in the stars. One practical difficulty is, that each of the heavenly bodies contains a number of different elements; so that in the light it sends us the hues arising from distinct substances are all blended into one beam. The first thing to be done is to get some way of splitting up a beam of light, so as to discover the components of which it is made. You might have a skein of silks of different hues all tangled together, and this would be like the sunbeam as we receive it in its unsorted condition. How shall we untangle the light from the sun or a star? I will show you by a simple experiment. Here is a beam from the electric light; beautifully white and bright, is it not? It looks so pure and simple, but yet that beam is composed of all sorts of colours mingled together, in such proportions as to form white light. I take a wedge-shaped piece of glass called a prism, and when I apply it in the course of the beam, you see the transformation that has taken place (Fig. 83). Instead of the white light you have now all the colours of the rainbow—red, orange, yellow, green, blue, indigo, violet, marked by their initial letters in the figure. These colours are very beautiful, but they are transient, for the moment we take away the prism they all unite again to form white light. You see what the prism has done: it has bent all the light in passing through it; but it is more effective in bending the blue light than the red light, and

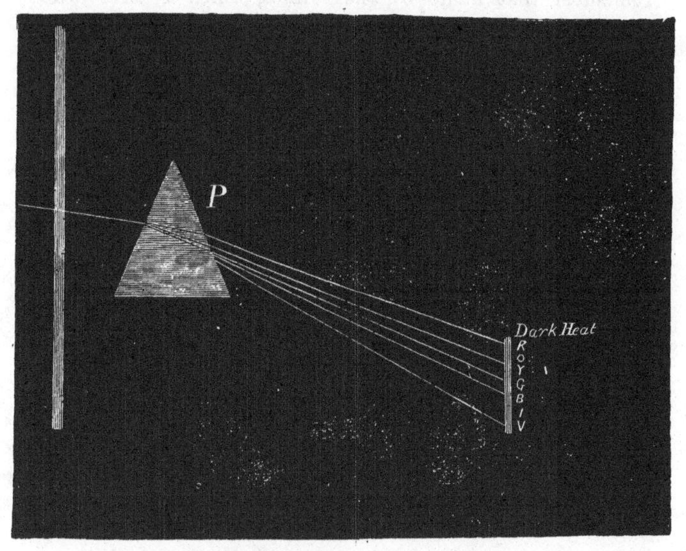

Fig. 83.—How to split up a Ray of Light.

consequently the blue is carried away much further than the red. Such is the way in which we study the composition of a heavenly body. We take a beam of its light, we pass it through a prism, and immediately it is separated into its components; then we compare what we find with the lights given by the different elements, and thus we are enabled to discover the substances which exist in the distant object whose light we have examined. I do not mean to say that the method is a simple one; all I am endeavouring to show is a general outline of the way in which we have discovered the materials present in the stars. The instrument that is employed for this purpose is called the spectroscope. And perhaps you may remember that name by these lines, which I have heard from an astronomical friend :—

> "Twinkle, twinkle little star,
> Now we find out what you are,
> When unto the midnight sky,
> We the spectroscope apply."

I am sure it will interest everybody to know that the elements the stars contain are not altogether different from those of which the earth is made. It is true there may be substances in the stars of which we know nothing here; but it is certain that many of the most common elements on the earth are present in the most distant bodies. I shall only mention one, the metal iron. That useful substance has been found in some of the stars which lie at almost incalculable distances from the earth.

THE NEBULÆ.

In drawing towards the close of these lectures I must say a few words about some dim and mysterious objects to which we have not yet alluded. They are what are called nebulæ, or little clouds; and they are justly called "little" clouds in one sense, for each of them occupies but a very small spot in the sky as compared with that which would be filled by an ordinary cloud in our air. The nebulæ are, however, objects of the most stupendous proportions. Were our earth and thousands of millions of bodies quite as big all put together, they would not be nearly so great as one of these nebulæ. Pictures of these objects show them to be like dull patches of light on the background of the sky. Astronomers reckon up the various nebulæ by thousands, but I must add that most of them are very small and uninteresting. A nebula is sometimes liable to be mistaken for a comet. The comet is, as I have already explained, at once distinguished by the fact that it is moving and changing its appearance from hour to hour, while scores of years may elapse without changes in the aspect or position of a nebula. The most powerful telescopes are employed in observing these faint objects. I take this opportunity of showing a picture of instruments suitable for such observations. It is the great reflector of the Paris Observatory (Fig. 84).

There are such multitudes of nebulæ that I can only show a few of the more remarkable kinds. In Fig. 85 will be seen pictures of a curious object in the constellation of Lyra seen under different telescopic powers. This object is a gigantic ring of luminous gas. To judge of the size of this ring let us suppose that a railway were laid across it, and the train you entered at one side was not to stop until it reached the other side, how long do you think this journey would require? I recollect some time ago a picture in *Punch* which showed a train about to start from London to Brighton, and the guard walking up and down announcing to the passengers the alarming fact that "this train stops nowhere." An old gentleman was seen vainly gesticulating out of the window and imploring to be

Fig. 84.—A Great Reflecting Telescope.

let out ere the frightful journey was commenced. In the nebular railway the passengers would almost require such a warning.

Let the train start at a speed of a mile a minute, you would think, surely, that it must soon cross the ring. But the minutes pass, an hour has elapsed; so the distance must be sixty miles, at all events. The hours creep on into days, the days advance into years, and still the train goes on. The years would lengthen out into centuries, and even when the train had been rushing on for a thousand years with an

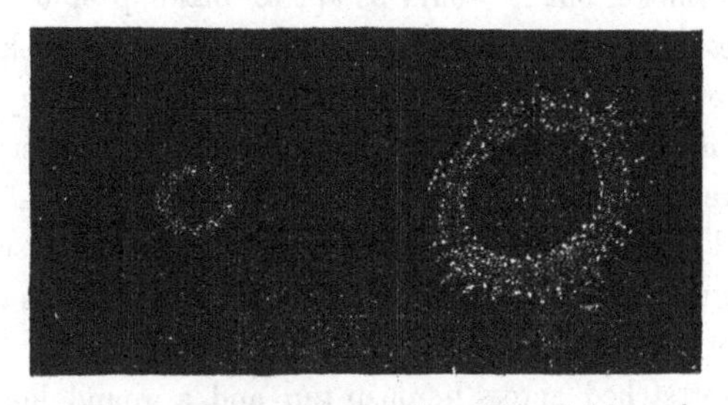

Fig. 85.—The Ring Nebula in Lyra, under different telescopic powers.

unabated speed of a mile a minute, the journey would certainly not have been completed. Nor do I venture to say what ages must elapse ere the terminus at the other side of the ring nebula would be reached.

A cluster of stars viewed in a small telescope will often look like a nebula, for the rays of the stars become blended together. A powerful telescope will, however, dispel the illusion and reveal the separate stars. It was, therefore, thought that all the nebulæ might be merely clusters so exceedingly remote that our mightiest instruments failed to resolve them into stars. But this is now known not to be the case. Many of these objects are really masses of glowing gas; such are, for instance, the ring nebulæ, of which I have just spoken, and the form of which I can simulate by a pretty experiment.

We take a large box with a round hole cut in one face, and a canvas back at the opposite side. I first fill this box with smoke, and there are different ways of doing so. Burning brown paper does not answer well, because the supply of smoke is too irregular and the paper itself is apt to blaze. A little bit of phosphorus set on fire yields copious smoke, but it would be apt to make people cough, and, besides, phosphorus is a dangerous thing to handle incautiously, and I do not want to suggest anything which might be productive of disaster if the experiment were repeated at home. A little wisp of hay, slightly damped and lighted, will safely yield a sufficient supply, and you need not have an elaborate box like this: any kind of old packing-case, or even a band-box, with a duster stretched across its open top and a round hole cut in the bottom, will answer capitally. While I have been speaking my assistant has kindly filled this box with smoke, and in order to have a sufficient supply, and one which shall be as little disagreeable as possible, he has mixed together the fumes of hydrochloric acid and ammonia from two retorts shown in Fig. 86. A still simpler way of doing the same thing is to put a little common salt in a saucer and pour over it a little oil of vitriol; this is put into the box, and over the floor of the box common smelling-salts is to be scattered. You see there are dense volumes of white smoke escaping from every corner of the box. I uncover the opening and give a push to the canvas, and you see a beautiful ring flying across the room, another ring and another follows. If you were near enough to feel the ring you would experience a little puff of wind; I can show this by blowing out a candle which is at the other end of the table. These rings are made by the air which goes into a sort of eddy as it passes through the hole. All the smoke does is to render the air visible. The smoke-ring is indeed quite elastic. If we send a second ring

Fig. 86.—How to make the Smoke-Rings.

hurriedly after the first we can produce a collision, and you see each of the two rings remains unbroken, though both are quivering from the effects of the blow. They are beautifully shown along the beam of the electric lamp, or, better still, along a sunbeam.

We can make many experiments with smoke-rings. Here, for instance, I take an empty box, so far as smoke is concerned, but air-rings can be driven forth from it, though you cannot see them, but you could feel them even at the other side of the room, and they will, as you see, blow out a candle. I can also shoot invisible air-rings at a column of smoke, and when the missile strikes the smoke it produces a little commotion and emerges on the other side, carrying with it enough of the smoke to render itself visible, while the solid black-looking ring of air is seen in the interior. Still more striking is another way of producing these rings, for I charge this box with ammonia, and the rings from it you cannot see. There is a column of the vapour of hydrochloric acid that also you cannot see; but when the invisible ring enters the invisible column, then a sudden union takes place between the vapour of the ammonia and the vapour of the hydrochloric acid: the result is a solid white substance in extremely fine dust which renders the ring instantly visible.

WHAT THE NEBULÆ ARE MADE OF.

There is a fundamental difference between these little rings that I have shown you and the great rings in the heavens. I had to illuminate our smoke with the help of the electric light, for, unless I had done so, you would not have been able to see them. This white substance formed by the union of ammonia and hydrochloric acid has, of course, no more light of its own than a piece of chalk; it requires other light falling upon it to make it visible. Were the ring nebula in Lyra composed of this material we could not see it. The sunlight which illuminates the planets might, of course, illuminate such an object as the ring if it were near us, comparatively speaking; but Lyra is at such a stupendous distance that any light which the sun could send out there would be just as feeble as the light we receive from a fixed star. Should we be able to show our smoke-rings, for instance, if, instead of having the electric light, I merely cut a hole in the ceiling and allowed the feeble twinkle of a star in the Great Bear to shine through? In a similar way the sunbeams would be utterly powerless to effect any illumination of objects in these stellar distances. If the sun were to be extinguished altogether the calamity would no doubt be a very dire one so far as we are concerned, but the effect on the other celestial bodies (moon and planets excepted) would be of the slightest possible description. All the stars of heaven would continue to shine as before. Not a point in one of the constellations would be altered, not a variation in the brightness, not a change in the hue of any star could be noticed. The thousands of nebulæ and clusters would be absolutely unaltered; in fact, the total extinction of our sun would be hardly remarked in the newspapers which are, perhaps, published in the Pleiades or in Orion. There might possibly be a little line somewhere in an odd corner to the effect that "Mr. So-and-So, our well-known astronomer, has noticed that a tiny star, inconspicuous to the eye, and absolutely of no importance whatever, has now become invisible."

If, therefore, it be not the sun which lights up this nebula, where else can be the source of its illumination? There can be no other star in the neighbourhood adequate to the purpose, for, of course, such an object would be brilliant to us if it were large enough and bright enough to impart sufficient illumination to the nebula. It would be absurd to say that you could see a man's face illuminated by a candle while the candle itself was too faint or too distant to be visible. The actual facts are, of course, the other way: the candle might be visible, when it was impossible to discern the face which it lighted.

Hence we learn that the ring nebula must shine by some light of its own, and now we have to consider how it can be possible for such material to be self-luminous. The light of a nebula does not seem to be like flame; it can, perhaps, be better represented by the pretty electrical experiment with Geissler's tubes. These are glass vessels of various shapes, and they are all very nearly empty, as you will understand when I tell you the way in which they have been prepared. A little gas was allowed into each tube, and then almost all the gas was taken out again, so that only a mere trace was left. I pass a current of electricity through these tubes, and now you see they are glowing with beautiful colours. The different gases give out lights of different hues, and the optician has exerted his skill so as to make the effect as beautiful as possible. The electricity, in passing through these tubes, heats the gas which they contain, and makes it glow; and just as this gas can, when heated sufficiently, give out light, so does the great nebula, which is a mass of gas poised in space, become visible in virtue of the heat which it contains.

We are not left quite in doubt as to the constitution of these gaseous nebulæ, for we can submit their light to the prism in the way I explained when we were speaking of the examination of the stars. Far from us as that ring in Lyra may be, it is interesting to learn that the ingredients from which it is made are not entirely different from substances we know on our earth. The water in this glass, and every drop of water, is formed by the union of two gases, of which one is hydrogen. This is an extremely light material, as you see by a little balloon which ascends so prettily when filled with it. Hydrogen also burns very readily, though the flame is almost invisible. When I blow a jet of oxygen through the hydrogen I produce a little flame capable of giving a very intense heat. For instance, I hold a steel pen in the flame, and it glows and sputters, and falls down in white-hot drops. It is needless to say that as a constituent of water hydrogen is one of the most important elements on this earth. It is, therefore,

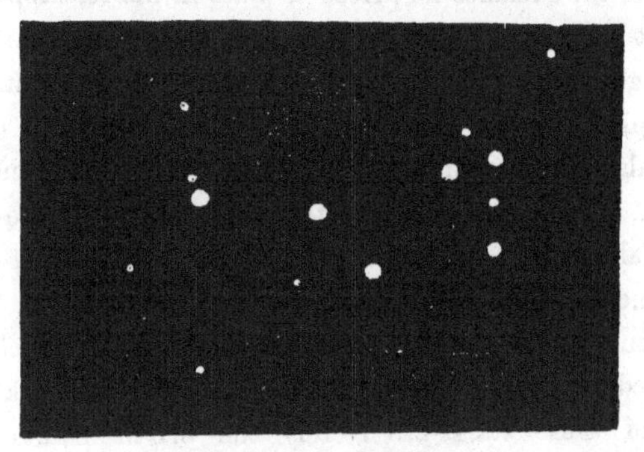

Fig. 87.—The Pleiades.

of interest to learn that hydrogen in some form or other is a constituent of the most distant objects in space that the telescope has revealed to us

PHOTOGRAPHING THE NEBULÆ.

Of late years we have learned a great deal about nebulæ, by the help which photography has given to us. Look at this group of stars which constitutes that beautiful little configuration known as the Pleiades (Fig. 87). It looks like a miniature representation of the Great Bear; in fact, it would be far more appropriate to call the Pleiades the

Little Bear than to apply that title to another quite different constellation, as has unfortunately been done. The Pleiades form a group containing six or seven stars visible to the ordinary eye, though persons endowed with exceptionally good vision can usually see a few more. In an opera glass the Pleiades becomes a beautiful spectacle; but in a very large telescope the stars appear too far apart to make a really effective cluster. When Mr. Roberts took a photograph of the Pleiades he placed a plate in his telescope, and on that plate the Pleiades engraved their picture with their own light. He left the plate exposed for hours, and on developing it not only were the stars seen, but there were also patches of faint light due to the presence of nebula. It could not be said that the objects on the plate were fallacious, for another photograph was taken, when the same appearances were reproduced.

When we look at that pretty group of stars, which has attracted admiration during all time, we are to think that some of those stars are merely the bright points in a vast nebula, invisible to our eyes or telescopes, though capable of recording its trace on the photographic plate. Does not this give us a greatly increased notion of the extent of the universe, when we reflect that by photography we are enabled to see much which the mightiest of telescopes had previously failed to disclose?

Of all the nebulæ, now numbering some thousands, there is but a single one which can be seen without a telescope. It is in the constellation of Andromeda, and on a clear dark night can just be seen with the unaided eye as a faint dull spot. It has happened, before now, that persons noticing this nebula for the first time, have thought they had discovered a comet. I would like you to try and find out this nebula for yourselves.

If you look at it with an opera-glass it appears like a

Fig. 88.—Taken from Mr. Roberts's Photograph of the Great Nebula in Andromeda.

dull spot, rather elongated. You can see more of its structure when you view it in larger instruments, but its nature was never made clear until some beautiful photographs were taken by Mr. Roberts (Fig. 88). Unfortunately, the nebula in Andromedra has not been placed conveniently for its portrait. It seems as if it were a rather flat-shaped object, turned nearly edgewise towards us. If you wanted to look at the pattern on a plate, you would naturally hold the plate square in front of you. You would not be able to see the pattern well if the plate were so tilted that the edge was turned towards you. That seems to be nearly the way in which we are forced to view the nebula in Andromeda. We have the same sort of difficulty about Saturn, we are never able to take a square look at his rings. We can trace in the photograph some divisions extending entirely round the nebula, showing that it seems to be formed of a series of rings; and there are some outlying portions which form part of the same system. Truly this is a marvellous object. It is impossible for us to form any conception of the true dimensions of this gigantic nebula, it is so far off that we have never yet been able to determine its distance. Indeed, I may take this opportunity of remarking that no astronomer has yet succeeded in ascertaining with accuracy the distance of any nebula, though everything points to the conclusion that they are at least as far as the stars.

It is almost impossible to apply the methods which we use in finding the distance of a star to the discovery of the distance of the nebulæ. These flimsy bodies are usually too ill-defined to admit of being measured with the precision and the delicacy required for the determination of distance. The necessary measurements can only be made from one star-like point to another similar point. If we could choose a star in the nebula and determine its distance, then, of course, we should have the distance of the nebula itself;

but the difficulty is that we have, in general, no means of knowing whether the star does actually lie in the nebula. It may, for anything we can tell, lie billions of miles nearer to us than the nebula, or billions of miles further off, and by merely happening to lie in the line of sight, appear to glimmer in the nebula itself.

If we have any assurance that the star is surrounded by nebula, then it may be possible to measure that nebula's distance. It will occasionally happen that grounds can be found for believing that a star which appears to be in the nebula does veritably lie therein, and is not merely seen in the same direction. There are hundreds of stars visible on a good drawing or a good photograph of the Andromeda nebula, and doubtless large numbers of these are merely stars which happen to lie in the same line of sight. The peculiar circumstances attending the history of one star, seem, however, to warrant us in making the assumption that it was certainly in the nebula. The history of this star is a remarkable one. It suddenly kindled from invisibility into brilliancy. How is a change so rapid in the lustre of a star to be accounted for? In a few days its brightness had undergone an extraordinary increase. Of course, this does not tell us for certain that the star lay in the nebula; but the most rational explanation that I have heard offered of this occurrence is that due, I believe, to my friend Mr. Monck. He has suggested that the sudden outbreak in brilliancy might be accounted for on the same principles as those by which we explain the ignition of meteors in our atmosphere. If a dark star, moving along with terrific speed through space, were suddenly to plunge into a dense region of the Great Nebula, heat and light would be evolved sufficiently to transform the star into a brilliant object. If, therefore, we knew the distance of this star at the time it was in Andromeda, we should, of course, learn the distance of the nebula. This has been attempted, and it has thus been proved that the Great Nebula must be very much further from us than is that star

Fig. 89.—To show how small is the Solar System in comparison with a great Nebula.

of whose distance I attempted some time ago to give you a notion.

We thus realise the enormous size of the Great Nebula. It appears that if, on a map of the nebula, we were to lay down, accurately to scale, a map of the solar system, putting the sun in the centre and all the planets around in their true proportions out to the boundary traced by Neptune, this area, vast as it is, would be a mere speck on the drawing of the object. Our system would have to be enormously bigger than it is if it were to cover anything like the area of the sky included in one of these great objects. Here is a sketch of a nebula (Fig. 89), and on it I have marked a dot which is to indicate our solar system. We may feel confident that the Great Nebula is as big if not bigger than this proportion would indicate.

CONCLUSION.

And now, my young friends, I am drawing near the close of that course of lectures which has occupied us, I hope you will think not unprofitably, for a portion of our Christmas holidays. We have spoken of the sun and of the moon, of comets and of stars, and I have frequently had occasion to allude to the relative position of our earth in the universe. No doubt it is a noble globe which we inhabit, but I have failed in my purpose if I have not shown you how insignificant is this earth when compared with the vast extent of some of the other bodies that abound in space. We have, however, been endowed with a feeling of curiosity which makes us long to know of things beyond the confines of our own earth. Astronomers can tell us a little, but too often only a little. They will say—That is a star, and That is a planet, and This is so big, and That so far; such is the meagre style of information with which we often have to be content, and, indeed, it is rare for us to receive even so much. The astronomers that live on other worlds, if their faculties be in any degree comparable with ours, must be similarly ignorant with regard to this earth. Inhabitants of our fellow-planets can know hardly anything more than that we live on a globe 8,000 miles across with many clouds around us. Some of the planets would not even pay us the compliment of recognising our existence; while from the other systems—the countless other systems—of space we are absolutely imperceptible and unknown. Out of all the millions of bodies which we can see, you could very nearly count on your fingers those from which our earth would be visible. This reflection is calculated to show us how vast must be the real extent of that universe around us. Here is our globe, with all its inhabitants, with its great continents, with its oceans, with its empires, its kingdoms, with its arts, its commerce, its literature, its sciences—all of which are naturally of such engrossing importance to us here—and yet it would seem that all these things are absolutely unknown to any inhabitants that may exist elsewhere. I do not think that any reasonable person will doubt that there are inhabitants elsewhere. There are millions of globes, many of them more splendid than ours. Surely it would be presumptuous to say that this is the only one of all the bodies in the universe, on the surface of which life, with all that life involves, is manifested? You will rather think that our globe is but one in the mighty fabric, and that other globes may teem with interest just as ours does. We can, of course, make no conjecture as to what the nature of the life may be elsewhere. Could a traveller visit some other globes and bring back to us specimens of the natural objects that he there found, no collections that the world has ever seen could rival them in interest. When I go into the British Natural History Museum and look around that marvellous collection it awakens in me a feeling of solemnity. I see there the remains of mighty extinct animals which once roamed over this earth; I see there objects which have been dredged from the bottom of the sea at a depth of some miles; there I can examine crystals which have required incalculable ages for their formation; and there I look at meteorites which have travelled from the heavens above down on to the earth beneath. Such sights, and the reflections they awaken, bring before us in an imposing manner the phenomena of our earth, and the extent and interest of its past history. Oliver Wendell Holmes said that the only way to see the British Museum was to take lodgings close by when you were a boy, and to stay in the Museum from nine to five every day until you were an old man: then you would begin to have some notion of what this Institution contains. Think what millions of British Museums would be required were the universe to be adequately illustrated: one museum for the earth, another for Mars, another for Venus—but it would be useless attempting to enumerate them!

Most of us must be content with acquiring the merest shred of information with regard even to our own earth. Perhaps a schoolboy will think it fortunate that we are so ignorant with respect to the celestial bodies. What an awful vista of lessons to be learned would open before his view, if only we had a competent knowledge of the other globes which surround us in space! I should like to illustrate the extent of the universe by following out this reflection a little further. I shall just ask you to join with me in making a little calculation as to the extent of the lessons you would have to learn if ever astronomers should succeed in discovering some of the things they want to know.

Of course, everybody learns geography and history. We must know the geography of the leading countries of the globe, and we must have some knowledge of their inhabitants and of their government, their resources and their civilisation. It would seem shockingly ignorant not to know something about China, or not to have some ideas on the subject of India or Egypt. The discovery of the New World also involves matters on which every boy and girl has to be instructed. Then, too, languages form, as also we know too well, an immense part of education. Supposing we were so far acquainted with the other globes scattered through space, that we were able to gain some adequate knowledge of their geography and natural history, of the creatures that inhabit them, of their different products and climates, then everybody would be anxious to learn those particulars; and even when the novelty had worn off it would still be right for us to know something about countries perhaps more populous than China, about nations more opulent than our own, about battles mightier than Waterloo, about animals and plants far stranger than any we have ever dreamt of. An outline of all such matters should, of course, be learned, and as the amount of information would be rather extensive, we will try to condense it as much as possible.

To aid us in realising the full magnificence of that scheme in the heavens of which we form a part, I shall venture to give an illustration. Let us attempt to form some slight conception of the number and of the bulk of the books which would be necessary for conveying an adequate description of that marvellous universe of stars which surround us. These stars being suns, and many of them being brighter and larger than our own sun, it is but reasonable to suppose that they may be attended by planetary systems. I do not say that we have any right to infer that such systems are like ours. It is not improbable that many of the suns around us have a much poorer retinue than that which dignifies our sun. On the other hand, it is just as likely that many of these other suns may be the centres of systems far more brilliant and interesting, with far greater diversity of structure, with far more intensity and variety of life and intelligence than are found in the system of which we form a part. It is only reasonable for us to suppose that, as our earth is an average planet, so our sun is an average star both in size and in the importance of its attendants. We may fairly take the number of stars in the sky at about one hundred millions; and thus we see that the books which are to contain a description of the entire universe — or rather, I should say, of the entire universe that we see—must describe 100,000,000 times as much as is contained in our single system. Of course, we know next to nothing of what the books should contain; but we can form some conjecture of the number of those books, and this is the notion to which I now ask your attention.

So vast is the field of knowledge that has to be traversed, that we should be obliged to compress our descriptions into the narrowest compass. We begin with a description of our earth; nearly all the books in all the libraries that exist at this moment are devoted to matters on or of this earth. They include all branches of history, all languages and literatures and religions, everything relating to life on this globe, to its history in past geological times, to its geography, to its politics, to every variety of manufacture and agriculture, and all the innumerable matters which concern our earth's inhabitants, past and present. But this tremendous body of knowledge must be immensely condensed before it would be small enough to retire to its just position in the great celestial library. I can only allow to the earth one volume of about 500 pages. Everything that has to be said about our earth must be packed within this compass. All terrestrial languages, histories, and sciences that cannot be included between its covers can find no other place on our shelves. I cannot spare any more room. Our celestial library will be big enough, as you shall presently see. I am claiming a good deal for our earth when I regard it as one of the most important bodies in the solar system. Of course it is not the biggest—very far from it; but it seems as if the big planets and the sun were not likely to be inhabited, so that if we allow one other volume to the rest of the solar system, it will perhaps be sufficient, though it must be admitted that Venus, of which we know next to nothing, except that it is as large as the earth, may also be quite as full of life and interest. Mars and Mercury are also among the planets with possible inhabitants. We are, therefore, restricting the importance of the solar system as much as possible, perhaps even too much, by only allowing it two volumes. Within those two volumes every conceivable thing about the entire solar system—sun, planets (great and small), moons, comets, and meteors—must be included, or else it will not be represented at all in the great celestial library. We shall deal on similar principles with the other systems through space. Each of the 100,000,000 stars will have two volumes allotted to it. Within the two volumes devoted to each star we must compress our description of the body itself and of the system which surrounds it; the planets, their inhabitants, histories, arts, sciences, and all other information. I am not, remember, discussing the contents, but only the number of the books we should have to read ere we could obtain even the merest outline of the true magnificence of the heavens. Let us try to form some estimate as to the kind of library that would be required to accommodate 200,000,000 volumes. I suppose a long straight hall, so lofty that there could be fifty shelves of books on each side. As you enter you look on the right hand and on the left, and you see it packed from floor to ceiling with volumes. We have arranged them according to the constellations. All the shelves in one part contain the volumes relating to the worlds in the Great Bear, while upon the other side may repose ranks upon ranks of volumes relating to the constellation of Orion.

I shall suppose that the volumes are each about an inch and a half thick, and as there are fifty shelves on each side, you will easily see that for each foot of its length the hall will accommodate 800 books. We can make a little calculation as to the length of this library, which, as we walk down through it, stretches out before us in a majestic corridor, with books, books everywhere. Let us continue our stroll, and as we pass by we find the shelves on both sides packed with their thousands of volumes; and we walk on and on, and still see no end to the vista that ever opens before us. In fact, no building that was ever yet constructed would hold this stupendous library. Let the hall begin on the furthest outskirts of the west of London, carry it through the heart of the City, and away to the utmost limits of the east—not a half of the entire books could be accommodated.

The mighty corridor would have to be fifty miles long, and to be packed from floor to ceiling with fifty shelves of books on each side, if it is to contain even this very inadequate description of the contents of the visible universe. Imagine the solemn feelings with which we should enter such a library, could it be created by some miracle! As we took down one of the volumes, with what mysterious awe should we open it, and read therein of some vast world which eye had never seen! There we might learn strange problems in philosophy, astonishing developments in natural history; with what breathless interest we should read of inhabitants of an organisation utterly unknown to our merely terrestrial experience! Notwithstanding the vast size of the library, the description of each globe would have to be very scanty. Thus, for instance, in the single book which referred to the earth I suppose just a little chapter might be spared to an island called England, and possibly a page or so to its capital, London. Similarly meagre would have to be the accounts of the other bodies in the universe; and yet, for this most inadequate of abstracts, a library fifty miles long, and lined closely with fifty shelves of books on each side, would be required!

Methuselah lived, we are told, nine hundred and sixty-nine years; but even if he had attained his thousandth birthday he would have had to read about 300 of these books through every day of his life before he accomplished the task of learning even the merest outline about the contents of space.

If, indeed, we were to have a competent knowledge of all these other globes, of all their countries, their geographies, their nations, their climates, their plants, their animals, their sciences, languages, arts, and literatures, it is not a volume, or a score of volumes, that would be required, but thousands of books would have to be devoted to the description of each world alone, just as thousands of volumes have been devoted to the affairs of this earth without exhausting the subjects of interest it presents. Hundreds of thousands of libraries, each as large as the British Museum, would not contain all that should be written, were we to have anything like a detailed description of the universe *which we see.* I specially emphasise the words just written, and I do so because the grandest thought of all, and that thought with which I conclude, brings before us the overwhelming extent of the unseen universe. Our telescopes can, no doubt, carry our vision to an immeasurable distance into the depths of space. But there are, doubtless, stars beyond the reach of our mightiest telescopes. There are stars so remote that they cannot record themselves on the most sensitive of photographic plates.

On the blackboard I draw a little circle, with a piece of chalk. I think of our earth as the centre, and this circle will mark for us the limit to which our greatest telescopes can sound. Every star which we see, or which the photographic plate sees, lies within this circle; but, are there no stars outside? It is true that we can never see them, but it is impossible to believe that space is utterly void and empty where it lies beyond the view of our telescopes. Are we to say that inside this circle stars, worlds, nebulæ, and clusters are crowded, and that outside there is nothing? Everything teaches us that this is not so.

We occasionally gain accession to our power by adding perhaps an inch to the diameter of our object-glass, or by erecting a telescope in an improved situation on a lofty mountain peak, or by procuring a photographic plate of increased sensibility. It thus happens that we are enabled to extend our vision a little further and to make this circle a little larger, and thus to add a little more to the known inside which has been won from the unknown outside. Whenever this is done we invariably find that the new region thus conquered is also densely filled with stars, with clusters, and with nebulæ; it is thus unreasonable to doubt that the rest of space also contains untold myriads of objects, even though they may never by any conceivable improvement in our instruments, be brought within the range of our observation. Reflect that this circle is comparatively small with respect to the space outside. It occupied but a small spot on this blackboard, the blackboard itself occupies only a small part of the end of the theatre, while the end of the theatre is an area very small compared with that of London, of England, of the world, of the solar system, of the actual distance of the stars. In a similar way the region of space which is open to our inspection, is an inconceivably small portion of the entire extent of space. The unknown outside is so much larger than the known inside, it is impossible to express the proportion. I write down unity in this corner and a cypher after it to make ten, and six cyphers again to make ten millions, and again, six cyphers more to make ten billions; but I might write six more, aye, I might cover the whole of this blackboard with cyphers, and even then I should not have got a number big enough to express how greatly the extent of the space we cannot see exceeds that of the space we can see. If, therefore, we admit the fact which no reasonable person can doubt, that this outside, this unknown, this unreachable, and, to us, invisible space does really contain worlds and systems as does this small portion of space in which we happen to be placed — then, indeed, we shall begin truly to comprehend the majesty of the universe. What figures are then to express the myriads of stars that should form a suitable population for a space inconceivably greater than that which contains 100,000,000 stars? But our imagination will extend still further. It brings before us these myriads of unseen stars with their associated worlds, it leads us to think that these worlds may be full to the brim with interests as great as those which exist on our world. When we remember that, for an adequate description of the worlds which we can see, one hundred thousand libraries, each greater than any library on earth, would be utterly insufficient, what conception are we to form when we now learn that even this would only amount to a description of an inconceivably small fragment of the entire universe?

Let us conceive that omniscience granted to us an adequate revelation of the ample glories of the heavens, both in that universe which we do see and in that infinitely greater universe which we do not see. Let a full inventory be made of all those innumerable worlds, with descriptions of their features and accounts of their inhabitants and their civilisations, their geology and their natural history, and all the boundless points of interest of every kind which a world in the sense in which we understand it does most naturally possess. Let these things be written every one, then may we say, with truth, that were this whole earth of ours covered with vast buildings, lined from floor to ceiling with book-shelves—were every one of these shelves stored full with volumes, yet, even then this library would be inadequate to receive the books that would be necessary to contain a description of the glories of the sidereal heavens.

CONCLUDING CHAPTER.

HOW TO NAME THE STARS.

EVERYONE who wishes to learn something about astronomy should make a determined effort to become acquainted with the principal constellations, and to find out the names of the brighter and more interesting stars. I have therefore added to STAR-LAND this little chapter, in which I have tried to make the study of the stars so simple, that by taking advantage of a few clear nights, there ought to be no difficulty in obtaining knowledge of a few constellations.

The first step is to become familiar with the Great Bear, or Ursa Major, as astronomers more generally call the group. We begin with this, because after it has been once recognised, then you will find it quite easy to make out the other constellations and stars. It may save you some trouble if you can get someone to point out to you the Great Bear, but even without such aid, I think you will be able to make out the seven bright stars which form this remarkable group, from the figure here given (Fig. 90). Of course, the position of this constellation, as of every other in the heavens, changes with the hour of the night, and changes with the seasons of the year. About April the constellation at 11 o'clock at night is high over your head. In September at the same hour, the Great Bear is low down in the north. It is to be seen in the west in July, and at Christmas it lies in the east at convenient hours in the evening for observation.

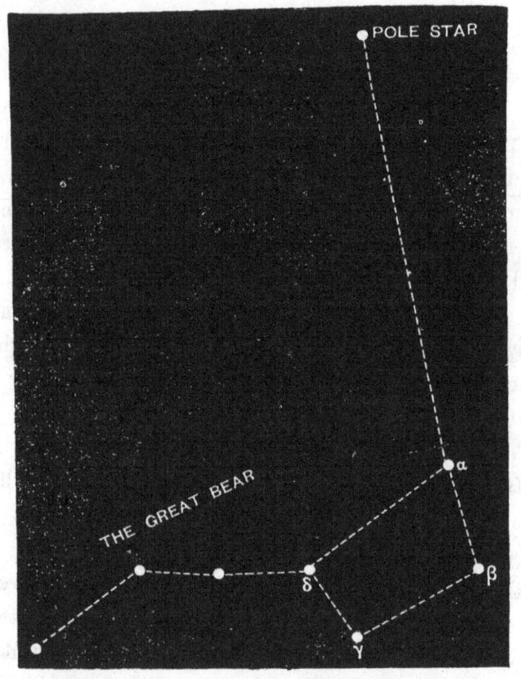

Fig. 90.--The Great Bear and the Pole Star.

One of the advantages from using the Great Bear as the foundation of our study of the stars, arises from the fact that to an observer in the British Islands or in similar latitudes this group never sets. Whenever the sky is clear after nightfall, the Great Bear is to be seen somewhere, while the brightness of its component stars makes it a conspicuous object. Indeed, there is only one constellation in the sky, namely, that of Orion, which is a more brilliant group than the Great Bear. We shall tell you about Orion presently, but it does not suit to begin with, because it can only be seen in winter, and is then placed very low down in the heavens.

Your next lesson will be to utilise the Great Bear for the purpose of pointing out the Pole Star. Look at the two stars marked α and β. They are called the "Pointers," because if you follow the direction they indicate along the dotted line in the figure, they will conduct your glance to the Pole. This is the most important star in the heavens to astronomers, because it happens to mark very nearly the position of the Pole on the sky. You will easily note the peculiarity of the Pole Star if you will look at it two or three times in the course of the night. It will appear to remain in the same place in the sky, while the other stars change their places from hour to hour. It is very fortunate that we have a star like this in the northern heavens; the astronomers in Australia or New Zealand can see no bright star lying near the Southern Pole which will answer the purposes that the Pole Star does so conveniently for us in the north.

The Pole Star belongs to a constellation which we call the Little Bear; two other conspicuous members of this group are the two "Guards;" you will see how they are situated from Fig. 80, p. 301. They lie nearly midway between the Pole Star and the last of the three stars which form the Great Bear's Tail. The same figure will also introduce us to another beautiful constellation, namely

Cassiopeia. You will never find any difficulty in identifying the figure that marks this group if you will notice that the Pole lies midway between it and the Great Bear.

Cassiopeia is also one of the constellations that never set to British observers; but now we have to speak of groups which do set, and which, therefore, can only be observed when the proper season comes round. The first

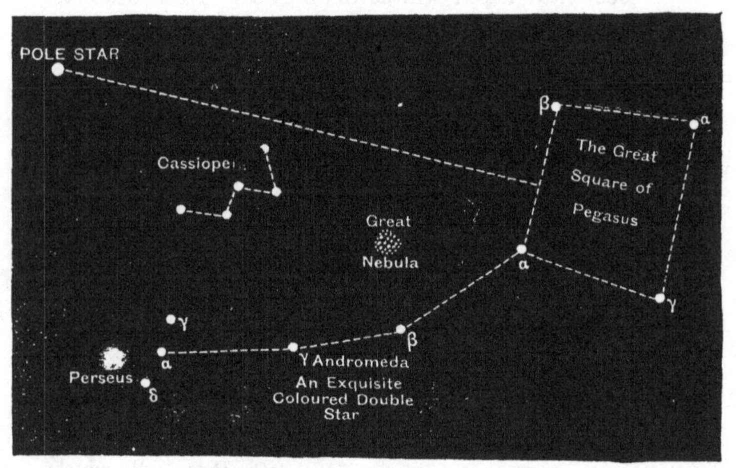

Fig. 91.—The Great Square of Pegasus.

of these is "the Great Square of Pegasus;" you cannot see this group conveniently in the spring or summer, but during the autumn and winter it is well placed after nightfall. There are four conspicuous stars forming the corners of the square, and then three others marked α, γ, and β (Fig. 91), which form a sort of handle to the square. In fact, if you once recognise this group, you will perhaps see in it a resemblance to a great saucepan with a somewhat bent handle, and then you will be acquainted with a large tract of Star-Land near the Square of Pegasus. From the figure you will see that a line imagined to be drawn from the Pole Star over the end of Cassiopeia, and then produced as far again, will just lead to the Great Square. I have also marked on this figure two objects that are of great telescopic interest; one of them is the Nebula in Andromeda, of which we had an account in the last lecture. You see it lies half-way between the corner α of the square and the group of Cassiopeia. Another interesting object is the star marked γ Andromedæ. The telescope shows it to consist of a pair of stars, the colours of which are beautifully contrasted.

At the end of this handle to the Great Square of Pegasus is the star α, in the constellation of Perseus. It lies between two other stars γ and δ. We refer to Fig. 82, in which these stars are shown. We there employed the figure to indicate the position of Algol, the remarkable variable one. Your map will also point out some other important stellar features. If we curve round the three marked γ, α, and δ of Pegasus, the eye is conducted to Capella, a gem of the first magnitude in the constellation of Auriga. Close to Capella is a long triangle, the corners of which are the "Hœdi," the three kids—which Capella is supposed to nurture.

If we carry a curve through γ, α, δ, of Perseus, and now bend it in the opposite way, the eye is led through ϵ and ζ in the same constellation, and then on to the Pleiades, of which we have already spoken.

Perseus lies in one of the richest parts of the heavens. The Milky Way stretches across the group, and the sky is strewn with stars beyond number. Even an opera-glass directed to this teeming constellation cannot fail to afford the observer a delightful glimpse of celestial scenery.

We may, however, specially remind the beginner that the objects on this map are not always to be seen, and as an illustration of the way in which the situation and the visibility of the constellations are affected by the time of year, I shall take the case of the Pleiades and follow them through a season. Let us suppose that we make a search for this group at 11 p.m. every night. On the 1st of January, the Pleiades will be found high up in the sky in the south-west. On the 1st of March, they will be setting in the west at the same hour. On the 1st of May, the Pleiades are not visible, neither are they on the 1st of July. On the 1st of September, they will be seen low down in the east. On the 1st of November, they will be high in the heavens in the south-east. On the ensuing 1st of January, the Pleiades will be found back in the same place which they occupied on the same date in the preceding year, and so on throughont the cycle. Of course, you will not suppose that their changes are due to actual motions in the group of stars themselves. They are merely apparent, and are to be explained by the motion of the earth round its axis, and the revolution around the sun.

Next we are to become acquainted with the glory of our winter skies, the constellation of Orion, Fig. 92. I dare say many of my readers are already familiar with the well-known twin stars which form the belt of Orion, but if not, they will be able to recognise it by the help of the groups already learned. Imagine a line drawn from the Pole Star through Capella, and then produced as much further

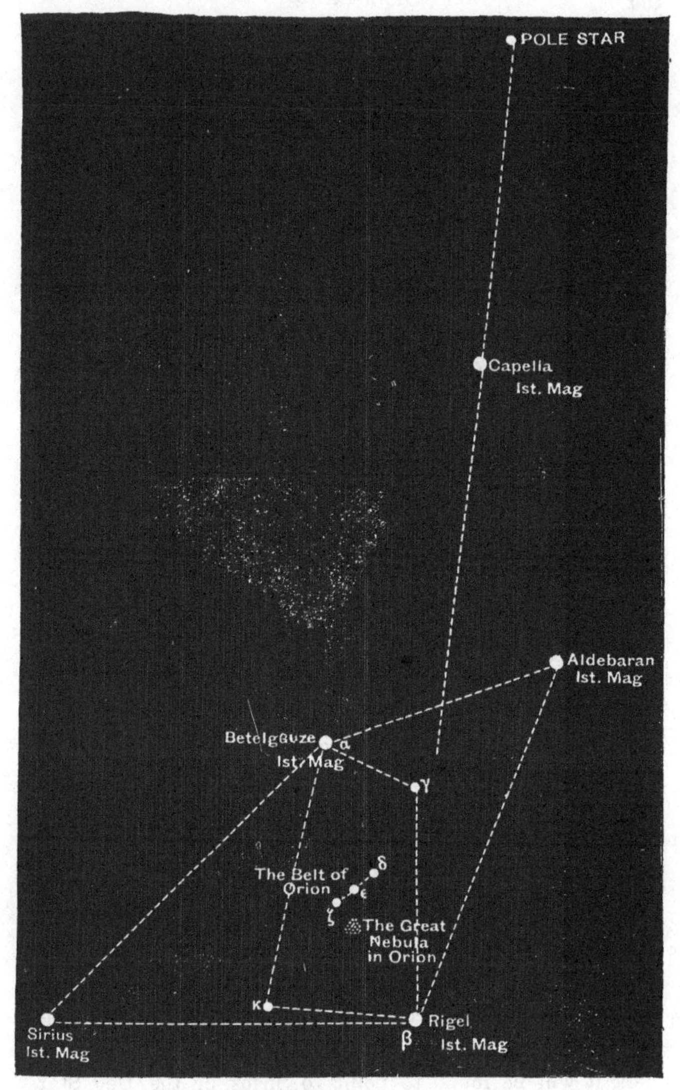

Fig. 92.—Orion and Sirius.

again, and we shall be conducted into the precincts of Orion. This group lies on the Equator, and, consequently, it is equally familiar to southern astronomers as to those of the north. It can be best seen by those who observe it from or near the Equator.

The brightest star in Orion is known either as α Orionis or as Betelgeuze, by which name it is represented in the figure. This star is of the first magnitude, and so is Rigel on the opposite side of the belt. The three stars of the belt and the two others, λ and κ, at which they point above and below, are of second magnitude.

The owner of a telescope finds especial attractions in this constellation. Notably among the objects which will interest him is the Great Nebula, the position of which is indicated in our figure. Under the middle of the belt are a few stars, around which is a hazy light that is perceptible with the smallest telescopic aid. Viewed by instruments of adequate proportions, these have developed into a marvellous nebula of glowing gas, attaining to dimensions so vast that no one has yet ever attempted to estimate them.

The vicinity of Orion is also enriched with some of the most interesting stellar objects. Follow the line of the belt upwards to the right, and your eye is conducted to a ruddy first magnitude star named Aldebaran, in the constellation of the Bull. This is a pleasing object, which the beginner will sometimes be apt to confuse with the planet Mars, to which, under certain circumstances, it certainly bears a resemblance. Another very pleasing little group, known as the Hyades, will be found near Aldebaran. If the line of the belt of Orion be carried down to the left, it will be found to point to Sirius, or the Dog Star.

You will find it an interesting occupation to make for yourself maps of small parts of the heavens. First copy out the chief stars in their proper places from the star atlas, and then fill in the smaller stars with your own observations. Try first on some limited region of the heavens; take the figure of Cassiopeia, for instance, or the Square of Pegasus, and see if you can produce a fair representation of those groups by marking in the stars that your instrument will show you; or take the Pleiades and make a tracing of the principal stars of the group from the sketch that we have given (Fig. 87), then take an opera-glass and fill in as carefully as you can all that it will show. I can assure you that you will find a little definite work of this kind full of interest and instruction.

But I hope you will desire to advance further in the study of the heavens. It is to be remembered that with even the most moderate of instruments there is much to be done. Many comets have been detected, and many planets have been discovered, by the use of telescopes so small that they could be easily carried out from the house for the evening's work and brought back again after the observations were over.

It remains for me to add a few words which will help you in finding the planets. It is, of course, impossible to represent such objects as Jupiter, Saturn, Venus, Mars, and Mercury on maps of the heavens, because these bodies are constantly moving about, and if their places were right to-day they would be wrong to-morrow. The almanac will be necessary for you here. You must find out by its help what planets are visible and in what part of the sky they are placed. Then you will have to compare your maps with the heavens, and when you find a bright star-like body that is not shown on your maps you may conclude at once that it is the planet. Although these objects are so star-like to the unaided eye, yet the resemblance is at once dispelled when we use a telescope. The star is only a bright point of light and white, the planet shows a visible shape. This is, at least, the case with the five planets I have named; for there are others, such as Uranus and Neptune, which are too far to be much more than star-like points in ordinary telescopes. The minor planets would not interest you.

I hope that the reading of STAR-LAND will, at all events, induce you to make a beginning of the study of the heavens, if you have not already done so. If you have the advantage of a telescope, so much the better; but, if this is impossible, make the best use of your own eyes. Do not put it off or wait till you get someone to teach you. If it be clear this very night, go out and find the Great Bear and the Pole Star, and as many of the other constellations as you can, and at once commence your career as an astronomer.

TABLE OF USEFUL ASTRONOMICAL FACTS.

THE sun's mean distance from the earth is 92,700,000 miles; his diameter is 865,000 miles, and he rotates in a period between 25 and 26 days.

The moon's mean distance from the earth is 238,000 miles; the diameter of the moon is 2,160 miles, and the time of revolution round the earth is 27·322 days.

THE PLANETS.

	Mean Distance from the Sun in Millions of Miles.	Periodic Time of Revolution in Days.	Diameter of Planet in Miles.	Axial Rotation.		
				Hrs.	Mins.	Secs.
Mercury ...	35·9	87·969	2,992	24	5	?
Venus	67·0	224·70	7,660	23	21	?
Earth	92·7	365·26	7,918	23	56	4·09
Mars	141	686·98	4,200	24	37	22·7
Jupiter ...	482	4,332·6	85,000	9	55	—
Saturn ...	884	10,759	71,000	10	14	23·8
Uranus ...	1,780	30,687	31,700	Unknown.		
Neptune ...	2,780	60,127	34,500	Unknown.		

THE SATELLITES OF MARS.

Name.	Mean Distance from Centre of Mars.	Periodic Time.		
		Hours.	Minutes.	Seconds.
Phobus...	5,800 miles.	7	39	14
Deimos...	14,500 ,,	30	17	54

THE SATELLITES OF JUPITER.

Name.	Mean Distance from Centre of Jupiter.	Periodic Time.			
		Days.	Hours.	Minutes.	Seconds.
I.	262,000 miles.	1	18	27	34
II....	417,000 ,,	3	13	13	42
III.	664,000 ,,	7	3	42	33
IV.	1,170,000 ,,	16	16	32	11

THE SATELLITES OF SATURN.

Name.	Mean Distance from Centre of Saturn.	Periodic Time.			
		Days.	Hours.	Minutes.	Seconds.
Ilenias	118,000 miles.	0	22	37	27·9
Enceladus	152,000 "	1	8	53	6·7
Tithys	188,000 "	1	21	18	25·7
Dione	241,000 "	2	17	41	8·9
Rhea	337,000 "	4	12	25	10·8
Titan	781,000 "	15	22	41	25·2
Hyperion	946,000 "	21	7	7	40·8
Japetus...	2,280,000 "	79	7	54	40·4

THE SATELLITES OF URANUS.

Name.	Mean Distance from Centre of Uranus.	Periodic Time. Days.
Ariel	119,000 miles.	2·520383
Umbriel	166,000 "	4·144181
Titania	272,000 "	8·705897
Oberon	363,000 "	13·463269

THE SATELLITE OF NEPTUNE.

Name.	Mean Distance from Centre of Neptune.	Periodic Time. Days.
Satellite	220,000 miles.	5·87690

THE END.

INDEX.

D.

E.

N.

O.

P.

Q.

R.

S.

T.

Printed in the United States
By Bookmasters